SpringerWienNewYork

Tort and Insurance Law
Vol. 20

Edited by the
European Centre of Tort
and Insurance Law

together with the

Research Unit for European Tort Law
of the Austrian Academy of Sciences

Saskia Klosse
Ton Hartlief (eds.)

Shifts in Compensating Work-Related Injuries and Diseases

With Contributions by

Esther F. D. Engelhard   Renaat I. R. Hoop
Ton Hartlief   Saskia Klosse
Niels J. Philipsen

SpringerWienNewYork

European Centre of Tort and Insurance Law
Landesgerichtsstraße 11
1080 Vienna, Austria
Tel.: +43 1 40127 1688
Fax: +43 1 40127 1685
E-Mail: ectil@ectil.org

Austrian Academy of Sciences
Research Unit for European Tort Law
Landesgerichtsstraße 11
1080 Vienna, Austria
Tel.: +43 1 40127 1687
Fax: +43 1 40127 1685
E-Mail: etl@oeaw.ac.at

This work is published with the financial support of
Netherlands Organisation for Scientific Research (NWO)

© 2007 Springer-Verlag/Wien
Printed in Germany

Springer-Verlag Wien New York is part of
Springer Science + Business Media
springer.at

Product Liability: The publisher can give no guarantee for the information
contained in this book. This also refers to that on drug dosage and application
thereof. In each individual case the respective user must check the accuracy
of the information given by consulting other pharmaceutical literature.

Typesetting: Composition & Design Services, Minsk, Belarus
Printing and binding: Strauss GmbH, 69509 Mörlenbach, Germany

Printed on acid-free and chlorine-free bleached paper
SPIN: 12032389

CIP data applied for

ISSN 1616-8623
ISBN 978-3-211-71555-0 SpringerWienNewYork

Table of Contents

Shifts in Compensating Work-Related Injuries and Diseases

S. Klosse and *T. Hartlief* (eds.)

1. Overall Framework: Shifts from Civil Law to Public Funding and vice versa

In 2003 the research programme *Compensation for damage – the shift from civil law to public funding and vice versa* started under the supervision of Prof. Dr Willem H. van Boom (Erasmus University Rotterdam) and Prof. Dr Michael F. Faure LL.M (University of Maastricht) and with financial support of the Dutch Scientific Organisation (NWO). Inspired by recent changes in the legal basis for compensating damage, this research programme aims at analysing shifts in paradigm with regard to compensation for damage across the borderline between civil law and public funding and vice versa. Several trends are discernable in this respect. On the one hand, there are clear signs of shifts being made from public funding to compensation systems based on civil law and individual responsibility. On the other hand, there are also developments which would seem to mirror shifts from individual responsibility to public funding or alternative systems of solidarity, such as compulsory insurance.

One of the key objectives of the research programme is to examine these shifts, and more in particular the reasons behind them. The programme does not have the ambition to provide a complete overview of all of these shifts. Instead, it concentrates on analysing shifts in four specific areas, notably in the field of work-related injuries and diseases, environmental damage, natural disasters and medical malpractice. Departing from the hypothesis that the reasons for a shift in these areas are diverse, the research programme reckons with the fact that a shift may have different effects on, for example, individual justice, operative efficiency, preventive incentives and insurability. Against this background, the shifts that can be identified in the four areas under review will be submitted to a thorough analysis in order to obtain a clear picture of their effects on the performances of a compensation system with regard to

* Prof. S. Klosse, Professor of Employment and Welfare Law, Faculty of Law, University of Maastricht.
** Prof. T. Hartlief, Professor of Private Law, Faculty of Law, University of Maastricht.

these particular aspects. To what extent do the shifts that can be distinguished make a positive contribution to, for example, individual justice and/or the prevention or the insurability of damage? Do they provide efficient incentives for individuals to protect themselves against the consequences of the facts of life on the one hand, and a just distribution of excessive burdens for society on the other hand? Are they, in other words, capable of striking a balance between private responsibility and public solidarity, and if so under which conditions? The ultimate goal of the research programme is to investigate whether it is possible to come up with an answer to these questions and to develop a framework on the basis of which both the performances of public and private compensation schemes and their interrelationship can be critically evaluated.

2.   Scope of this Book

3   This book can be seen as a result of the first stage of the research programme which aimed at spotting and evaluating shifts in the four research areas. The findings of these surveys will be dealt with in separate books. This volume provides a case study on developments that occurred overtime in compensation systems for damage arising from work-related injuries and diseases. The book intends to contribute to gaining a deeper understanding of these developments, thereby focussing on shifts that took place in the compensation systems for work-related injuries and diseases in Germany, England, Belgium and the Netherlands. The shifts that can be observed in the legal doctrine, legislation and the case law of these countries will first be mapped from a historical and comparative perspective. Within this context, attempts will be made to uncover the precise nature of the given shifts: Do they imply a transition from private law to public funding and if so to what extent? And do the shifts that can be spotted in the legal systems under review occur in the same direction or are contradicting trends discernable? Subsequently efforts will be made to analyse and explain the shifts on the basis of the legal history so as to comprehend the reasons behind the shifts that can be identified in the four countries under review. What did the legislator expect to achieve and why? What were the driving forces to make certain alternations and to what extent were these alterations inspired by various interest groups? Next to this primarily theoretical and legal analysis, this case study will examine the effects of the shifts on, for example, the level of compensation and deterrence of damage on the basis of the available empirical data. Did the changes that have been made in the legal systems under review have the desired effects on the compensation or deterrence of work-related personal damage or did other, perhaps unexpected, effects arise and if so why and in what form? Since a merely legal approach will not suffice to provide an answer to these questions, an economic analysis of law will be included to analyse the outcome of the shifts in the four countries under review.

4   In view of the complex and rather complicated nature of the questions that are to be answered, it is not possible to address them in full. For this reason, their content will be touched upon only from some angles and with respect to a few

legal systems. Within this context the contributors to this volume, Esther Engelhard, University of Utrecht (chapter I), Renaat Hoop, University of Maastricht and Brussels (chapter II) and Niels Philipsen, University of Maastricht (chapter III and IV), will focus their attention on uncovering relevant shifts and explaining the motives behind them. It is not their ambition to include normative statements, for example, on the question of whether work-related damage should ideally be compensated through public funding or rather through a private law mechanism or a mix of both. In order to do so, far more (also empirical) research would be necessary which would exceed the scope of this book.

3. What Is a Shift?

In order to cover damage caused by work-related injuries and diseases, differ- 5
ent compensation schemes have been developed in the course of time. These schemes vary from systems primarily based on civil liability to systems based on (social) insurance but mostly a combination of both. Over the years these systems have been subject to many changes. However, not all of these changes brought about shifts in paradigm with regard to the compensation of damage caused by work-related injuries and diseases. To be characterised as such more precise criteria are needed. In this book a change in an existing compensation mechanism is regarded as a shift when a new policy is introduced that seeks to alter the way in which damage resulting from work-related injuries and diseases is to be compensated by rearranging the legal obligation to bear the damage. Hence, a shift in the meaning of this book implies a reallocation of losses through a change within or between existing compensation systems. Such a change may reveal itself in a conversion of the legal basis for compensating damage from public to private law or the other way round, but may also manifest itself in a transition from, for example, voluntary to compulsory insurance or from civil liability based on fault to strict liability. From the chosen definition to identify a shift it follows that a change within a compensation system that merely alters the conditions to claim compensation, for example by mitigating the requirement of the damage being purely work-related or by tightening the eligibility criteria for claiming benefits or lowering the benefit level, will not be regarded as a shift in the meaning of this book, since such a change does not rearrange the legal basis to bear the damage.

4. Overview of the Shifts in Compensation in the Field of Work-Related Injuries and Diseases

In the first chapter of this book, Esther Engelhard provides an overview of the 6
shifts within or between compensation systems that have occurred since the end of the 19th century in Germany, England, Belgium and the Netherlands. Engelhard's chapter shows that these four countries share a long history of legislative endeavours to regulate the employee's right to compensation for work-related personal damage, mostly through a combination of civil liability and social or private insurance. Yet, different approaches can be observed towards the mutual relationship between these compensation systems. At the

end of the 19th century, for example, Germany and the Netherlands shifted from a compensation system primarily based on civil liability to a social insurance system with a collective and compulsory character. In this new system the possibility to claim compensation for work-related injuries and diseases on the basis of civil liability was largely put aside. Belgium and England on the other hand, remained closer to the existing system primarily based on liability of the employer. Nevertheless it is possible to spot a shift within this system, since strict liability for employers was introduced so as to make the system more easily accessible for injured employees.

7   After the Second World War, the wish to establish an all-embracing social security system became predominant. In Belgium and England, this development resulted in combining strict liability with a legal obligation for employers to take out insurance for the risk of being held liable for work-related injuries. In Belgium this change was regarded as a shift towards a social insurance system. In Germany, the new sense of solidarity brought about by the war, inspired the legislator to broaden the scope of the already existing social insurance for work-related injuries and diseases by alleviating the requirement of causality between the damage and work-related activities. In the Netherlands a more fundamental change was made. There, the existing social insurance for work-related injuries and diseases was integrated in a new mandatory social insurance system which no longer made a distinction between work-related and non-work related injuries and diseases. This shift was accompanied by a reintroduction of the possibility to claim compensation for work-related damage on the basis of civil liability of the employer. Thus civil liability revived as a complementary scheme so as to make full compensation possible. Unlike in Belgium and Germany, the possibility to undertake legal action had also been preserved in England as a complementary scheme to obtain full compensation.

8   Over the years, an increase in the use of this complementary system can be observed in both England and the Netherlands. In both countries it would seem that this trend has been induced by two other developments, notably the felt need to cut public expenditure, especially on social insurance benefits, and an increased claim-consciousness of victims of work-related personal damage. In the Netherlands (and in England to a lesser extent), this trend has been accompanied by a call for reduced government intervention and a warm enthusiasm for different forms of privatisation so as to strengthen the responsibility of employers and employees to prevent damage arising from work-related injuries and diseases. This trend would seem to imply a shift back from public funding towards private responsibility and a more significant role of private funding of losses through private insurance and civil action based on liability of the employer.

5. Explanation of the Reasons for the Shifts Within or Between Compensation Systems

In the second chapter, Renaat Hoop elaborates on the findings of the first   9
chapter, thereby concentrating on the motives behind the shifts that have taken
place over the years in the four countries under review. The chapter reveals
what arguments have been put forward in favour or against a proposed shift
during the parliamentary proceedings and by advisory committees and also
sheds light on what policymakers and the legislator did expect to achieve or
gain with a certain shift within or between the existing compensation systems.
Thus, Hoop provides insight in both the purpose of a shift and the extent to
which it was meant to make a positive contribution to, for example, individual
justice, operative efficiency, preventive incentives or the insurability of damage caused by work-related injuries and diseases.

Hoop's chapter shows that, all in all, it would seem justifiable to conclude that   10
the majority of the shifts that came about in the four countries under review
have been inspired by the ambition to enhance the notion of individual justice
by refining and broadening the possibilities to claim compensation for work-
related damage. At the end of the 19th century, for example, the social misery
caused by the industrialisation combined with pressure from the social movement brought governments to profoundly question the adequacy of civil liability as the dominant compensation scheme. Awareness grew that many victims
of work-related injuries and diseases lived in poverty due to deficiencies of
this scheme which was costly and slow and put a heavy burden of proof on injured employees. The main motive to shift to another compensation system
was to improve this situation by replacing the existing system by a system that
was more easily accessible and offered a more adequate compensation to victims of work-related injuries. After the Second World War, the aspiration to
establish a comprehensive social security system, which was stirred by a new
sense of solidarity stemming from the war, strengthened the drive to improve
and expand the compensation facilities for victims suffering from work-related personal damage in the four countries under review. Again, this development was basically motivated by the ambition to adequately respond to the notion of individual justice.

The developments that took place in England would seem to be the only ex-   11
ception to this rule, since in England the pursuit for safer working conditions
has also been an important source of inspiration from the start. For this reason,
the English legislator initially rejected the principle of compulsory insurance
as this would be counterproductive to preventive efforts of the employer. After
the Second World War the principle of compulsory insurance was accepted
which pushed the importance of providing the system with preventive incentives to the back seat. Recently this objective gained ground again as a result
of the growing costs of work-related injuries and diseases. In order to keep
these costs within limits attempts have been made to partially transfer these
costs to the employer and the victims of work-related personal damage. In the

Netherlands a comparable development can be observed. In this country, employers have been more directly confronted with the costs of sick-leave and incapacity work, for example, by making employers responsible for bearing these costs during the first two years of incapacity to work due to illness. Compared to the other countries under review, this is a heavy burden, particularly in view of the fact that the Dutch system does not make a distinction between work-related and non work-related injuries and diseases. The idea behind this form of privatisation is to give employers and employees a strong incentive to take adequate measures to curb the risk of sick-leave and incapacity to work. Expectations are that this development will not only have a positive preventive effect, but will also enhance operative efficiency since it shifts the responsibility to design effective preventative policies to the company level where the problems related to sick-leave and incapacity to work can be tackled more directly.

6.   *Empirical Analysis*

12   In the third chapter of the book, Niels Philipsen presents empirical data which have been collected in order to gain an impression of the effectiveness of the shifts that have been identified in the previous chapters. Within this context, Philipsen examined, for example, if and to what extent it would be possible to draw conclusions from the available data on the deterrence effect of certain shifts. Do the available empirical data, for instance, point to a reduction of the number of work-related injuries and diseases or to a decrease in the number of claims for compensation for this type of damage after a certain shift has been made? Philipsen's chapter shows that it is hardly feasible to come up with a solid answer to this question. In essence, this can be attributed to the fact that relevant data to back up the findings of the previous chapters have often become available only fairly recently or, particularly when insurers are involved, are not available at all, for example, for reasons of confidentiality or competitiveness. Consequently, the findings in this chapter are, from sheer necessity, based on more general indicators, such as the (estimated) number of work-related injuries and diseases (if possible including developments over time related to that number) and general data on compensation systems based on either social security or tort law or private insurance in so far as these systems apply to work-related injuries and diseases. However, it is difficult to interpret these data due to differences in the definitions used and to changes that have been made in the way of reporting and presenting the data.

13   In spite of this, Philipsen's chapter displays that it is possible to identify certain trends. In England, for example, the rate of major work-related injuries has risen recently, which would seem to be caused by more reported accidents in the service sector. In Belgium, England and Germany, on the other hand, there is a downward trend discernable in the general rates for work-related injuries in recent years. Unfortunately, a clear explanation for this trend is hard to give, mainly as a result of under-reporting. For this reason it remains uncertain to what extent this downward trend can be ascribed to successful preven-

tive efforts. Only in Germany, where the decrease in work-related diseases is most apparent, preventive measures would seem to have had a positive effect, next to changes in the economic structure.

Philipsen's chapter also provides information on the amounts paid out to vic-   14
tims of work-related injuries and diseases by either insurers or funds. On the basis of this information, it is only possible to conclude that, in general, the average amount that has been paid out to victims has risen over the years. It would seem justifiable to attribute this trend to the increased costs of serious injuries. However, more information is needed to perform a further analysis.

7.   *Empirical Findings in the US*

In view of the difficulties that Philipsen encountered in finding relevant data   15
on the basis of which the effectiveness of certain shifts can be evaluated, it is interesting to take a look at the situation in the US where similar develop-ments have taken place and, contrary to the situation in Europe, relevant data are available on the basis of which some effects of these developments can be assessed. Against this background, Philipsen intends to complete the picture in the fourth chapter of this book by providing some valuable empirical back-ground information on, for example, the deterrence and compensation effects of tort law, risk-related insurance premiums, no fault compensations systems and safety regulation, that can be derived from empirical research conducted in the US.

The chapter starts with clarifying some relevant law and economic aspects,   16
thereby highlighting the ongoing debate in the American literature. Then it goes more deeply into the existing empirical literature on the shift from civil liability to workers' compensation based on compulsory insurance that has oc-curred in the US. Unlike in Europe, there is highly interesting literature avail-able in the US, which specifically deals with both the effects of this shift on the deterrence and compensation of work-related injuries and the impact of health and safety regulations and risk-related premiums on, for example, ben-efit levels and employers' costs. On the basis of this literature, it would seem justifiable to conclude that the shift from tort law towards workers' compensa-tion based on compulsory insurance has had a slightly positive effect on the deterrence of work-related injuries. However, it should be noted that this does not apply to work-related diseases. Moreover, it should be kept in mind that the workers' compensation system is often intertwined with other compensa-tion mechanism, such as social and private insurance. For this reason, it is quite hard to measure the performance of the workers' compensation scheme as far as compensation is concerned. Another observation that can be derived from the American literature is that the impact of health and safety regulations on the deterrence of work-related personal damage would seem to be low, mainly because of enforcement problems. The experience rating in the work-ers' compensation scheme, on the other hand, has, in general, been found to be more effective. All in all, Philipsen deduces from the American literature that,

as in Europe, the debate on the pros and cons of the workers' compensation scheme is far from over. The book winds up with a few concluding observations of the editors: Ton Hartlief and Saskia Klosse.

8.   *Words of Thanks*

17   We are grateful for the willingness of all those who contributed to the preparation of this book on the basis of the pattern and the questions provided. Equally grateful are we to the general coordinators of the research programme, Prof. Dr. Willem van Boom and Prof. Dr. Michael Faure, for their support in conducting this case study on shifts in the compensation of work-related personal damage. We owe also gratitude to the Dutch Scientific Research Council (NWO) for the financial support this organisation provided to the research programme *Compensation for damage – the shift from civil law to public funding and vice versa* of which this case study is a part. In addition, we are grateful to the European Centre for Tort and Insurance Law (ECTIL) through which our contacts with the publisher have been arranged and which edits the series in which this book is published. More in particular, we owe gratitude to Mrs Stockenhuber of ECTIL for reviewing the texts, the footnotes and the references before publication, and to Mrs. Steininger for her coordinating activities. Finally, we owe thanks to Mrs. Kuijpers of the Maastricht European Institute for Transnational Legal Research (METRO) for editorial assistance.

The text of this volume has been put in its final form in June 2006. Developments that took place after that date could therefore not be taken into account.

Maastricht, June 2006   Saskia Klosse
                                                                              Ton Hartlief

Shifts of Work-Related Injury Compensation. Background Analysis: The Concurrence of Compensation Schemes

*E.F.D. Engelhard**

1. Introduction

One morning in the early fall of 1996 Ms. Peters, in her daily routine as a   1
member of the cleaning staff of Hofkens Ltd., is making an effort to remove a
coffee stain off a desk leg in one of the offices. Unfortunately as she is bend-
ing forward, leaning with her one hand on the desk while cleaning with the
other, she loses balance, trips and hurts her finger, causing muscular dystrophy
and permanently disabling her from working at full speed.[1] A year later Mr.
Dusarduyn, an experienced roofer who is told by his employer to go on the
roof and fix it falls through a hole covered with insulation roofing material almost three metres down and brakes his wrist.[2] In both of these cases and their
subsequent proceedings damages were claimed from the victim's private employer. Both employees, Peters and Dusarduyn, argued to have suffered income loss for having been unable to work for some time, the latter even
claimed to have lost profits, as – in addition to his contract of services – he
was part-time self-employed. Furthermore each argued to have incurred medical expenses and suffered non pecuniary damages.

Aside from the legal proceedings over these matters more fundamental
questions are raised by such claims regarding the (concurrence of) systems
that may provide compensation for the damages of the victim-employee. Who
will have to pay for these losses and why?

In Germany, Belgium, England, the Netherlands and many other European le-   2
gal systems the recovery of damages caused by workplace injuries is to a great

---

* University of Utrecht, Molengraaff Institute for Private Law, The Netherlands, E-Mail: e.engel-
hard@law.uu.nl. This research was part of NWO's 'Shifts of Governance' project. The author
would like to thank the law faculties of both the University of Maastricht and the University of
Utrecht where this research was conducted.
[1] Example taken from a Dutch Supreme Court's decision on the employer's civil liability in this
instance, HR 12 September 2003, NJ 2004, 177 (*Peters/Hofkens BV*).
[2] HR 16 May 2003, *NJ* 2004, 176 (*Dusarduyn/Du Puy BV*).

extent governed by statutory rules on public and/or private insurance and liability law. All seem to share in common a long history of legislative endeavours to regulate the employee's right to compensation for workplace accidents. Also in each country several of these public and private law systems are
in operation, therefore each has different 'layers' of protection, as Philipsen
rightly calls them in his contribution. But the differences between these rules
if we restrict ourselves, as we will do here, to the four law systems as just
mentioned are enormous. In the first three the main source of compensation
lies in basic mandatory accidents insurance, the first 'layer' of compensation
so to speak, albeit with great differences regarding the level of compensation
and the availability of additional sources of compensation such as liability
law. The opposite legal approach is taken by the Netherlands, as Dutch basic
compensation comes from social security rules that make no difference between accidents caused at work or otherwise. However here, special protection is offered by a 'second layer of protection', to stay with Philipsens way of
putting it: a rather strict – de facto 'no-fault' oriented – regime of tort law liability for damages that are left uncompensated by social security. That means
that where social security cuts reduce the level of basic protection, tort law
must come to the fore by offering additional protection.

3   At this point we are touching upon our main interest here: In what ways do
      shifts of policy in one compensation system have an effect on other compensa-
      tion systems? While Philipsen in his contribution addresses this question from
      an economic perspective, I will explain how the various compensation
      schemes in each law system work by discussing the main rules on recovery
      and on the concurrency of resources and how they have developed through the
      years. This will be done for Dutch law successively followed by a comparison
      with British[3], German and Belgian law[4]. Inevitably, in doing so, we will touch
      upon the main purposes and rationale of the relevant rules and provisions; for
      further discussion of this aspect, see the contribution by Hoop. The focus will
      be on the compensation of regular employees – not on state officials, indepen-
      dent contractors or entrepreneurs – and on industrial (workplace) accidents,
      rather than on occupational diseases.
         The main questions to be answered for each law system are: What roles do
      private or public insurance and tort law play with regard to recoveries by em-
      ployees and how do these separate systems interact (in the sense that changes
      of the level of compensation or the grounds for recovery in one system will
      have ramifications for the other)? And to what extent are employers expected
      to bear the financial risk of damages caused at work?

---

[3] Hereafter reference will be made to 'England' instead of 'Great Britain', although most social
    security legislation is made applicable to 'Great-Britain' (and the Northern-Irish social legisla-
    tion seems to be similar for the better part).
[4] For a brief comparative overview in Dutch of the compensation of occupational *disease* see
    H.J.W. van Dongen, in: M. Faure/T. Hartlief (eds.), *Verzekering en de groeiende aansprakelijk-
    heidslast* (1995), 87 *et seq.*

2. What Defines a 'Shift in Governance'?

Surely to analyse policy changes or better yet 'shifts in governance' one must   4
define what it is that needs to be changed. My main focus will be on *any pre-empted way of compensating damages of the employee (or his relatives)* caused by workplace accidents or occupational diseases (as opposed to factual 'compensation' through winning the lottery for instance). As I see it, and in line I think with the explanation given in the introduction by Hartlief and Klosse, the term 'change' or 'shift' as used below means: any – historical or future planned – change within or between compensation systems to the effect that someone other than before must bear the damage. The 'shifts in governance' referred to below therefore seek to alter the manner in which damages caused by workplace injuries must be compensated.

I have found three dimensions of change to be relevant here. First, the change must have *legal effect* in the sense that it affects the legal position of the employee (or his relatives) and/or of the natural or legal persons responsible for his (their) damages. An example of this is when social security cuts lead to the effect that the victim-employee will not be entitled to as many benefits for his losses than he was before, for which he will then have an action for damages. Secondly, the focus will be on changes which take place on a central level, in the sense that they affect private individuals or public/private law institutions *in general* (as opposed to concrete case decisions which – at least primarily – merely affect the individual parties involved). Last and most importantly, this new policy must render a *reallocation of losses*, which means that it must rearrange the legal obligation to compensate in order to make someone else pay. My analysis will thus be limited to legislative decisions that in effect favour one compensation system over the other with regard to damages caused by the so-called *'risque professionnel'* (the risk of getting injured or any other detriment due to work).

3. Shifts that Matter: A Three-Part Overview

There are abundantly many examples of changes in governance involving the   5
allocation of losses, especially in the sphere of social security law.[5] For convenience of comparison I consider it practical to split the relevant changes within in the four law systems into three significant periods of time.

The first period of time starts I think in the late 19th century when – private law – workmen's compensation schemes first took form, covering the risk of damages caused by workplace accidents. Secondly throughout the beginning and halfway through the twentieth century the introduction of social security plans forms a significant change in all four law systems. Protection was considered to be a 'right' and was extended to all risks regarding illness, injury or death – regardless of the exact cause. As I see it this development came more or less to a stop with the need for social reform in the 1970s. From that

---

[5] Again, one might think of social security cuts, for example with regard to benefits for financial dependants in case of death, to the effect that private individuals who seek protection must take out private insurance.

point onwards the third time period starts, which can be characterized by its market oriented approach promoting personal responsibility and a differentiation of risks.

Following this timeframe the relevant shifts of each of the four systems will now be discussed. It should be noted that my – rather random – choice regarding the exact years in which each period of time starts and ends, merely serves to keep an orderly discussion.

4. Shifts I (1870–1920): From Civil Liability to Private Insurance

4.1 Introduction

6 The political debate regarding the recovery of employees suffering from industrial injuries in Western Europe was initiated in the late eighteenth century. For financial support employees who suffered from an injury generally had to rely on their families, private initiatives for poor relief or on guilds, later in time followed by collective corporation funds such as the *Friendly Society* in England, and finally commercial insurance. Clearly the individual employee's social dependency had been worsened by modern industry with its size, its complexity and impersonal nature, built on *laissez-faire* notions and freedom of contract.[6] Little attention was paid to occupational health and safety or industrial hazards. Only through massive organizations could one hope to bargain on equal terms with the new management. In theory those who could afford it had a system of civil liability to their avail, yet actions *vis-à-vis* the employer were often ridden by serious financial, legal and/or social barriers.[7] Clearly this created a growing welfare concern in West-European countries and the need for protection against the dangers of life and body that most employees and their families had to endure. But also the (liberal) establishment in most of these countries feared the social movement which was slowly setting foot in the 1840s[8].

---

[6] Th. Nipperdey, *Deutsche Geschichte 1866–1918*, Erster Band. Arbeitswelt und Bürgergeist (1998), 335: 'Das Neue gegenüber alten Armutsproblemen war, dass weder Familie, Kirche und Gemeinde noch karitative Aktivitäten oder die Bemühungen paternalistischer Unternehmer das bewältigen konnten'.

[7] P. Cohen, *The British System of Social Insurance* (1932) who stresses the poor position of the worker's financial dependants in case of a fatal accident; D.G. Hanes, *The First British Workmen's Compensation Act 1897* (1968), 6 ('The common law, as an instrument of relief, was altogether inadequate and fraught with difficulties; had it not been so, there would be no need for the step eventually taken'.); H. Barta, *Kausalität im Sozialrecht. Entstehung und Funktion der sogenannten Theorie der wesentlichen Bedingung* (1983); L. Bier, *Aansprakelijkheid voor bedrijfsongevallen en beroepsziekten* (1988), 114 *et seq.*; R.J.S. Schwitters, *De risico's van de arbeid. Het ontstaan van de Ongevallenwet 1901 in sociologisch perspectief* (1991), 90.

[8] H. Barta (*supra* fn. 7), 51.

4.2 Germany's Insurance Model of Workers' Compensation (1870–1920)

4.2.1 *From Prussian Fault Liability to Bismarck's Reichshaftpflichtgesetz (Liability Act, RHpflG) 1871*

The 1871 German empire as founded by Bismarck seems central in this re- 7
spect as it was first in launching programs to institute social welfare schemes
for employees. These initiatives were rooted in the 'social cause', the rising
voice of industrial workers regarding their poor socio-economic position.[9] A
growing number of industrial entrepreneurs had started to recognize the value
of their labour force and the need to bring social peace by protecting it from
major life risks such as disability and retirement. Local health, disability and
retirement plans were introduced and the tradesmen's guilds had been requir-
ing membership in such plans. On a national level the idea started to set foot to
develop uniformly regulated public law schemes for health insurance, protection against old age and disability *and industrial accidents insurance*. According to more recent historical studies[10], Bismarck's ideas had already come
about in the early 1870s, inspired by – nationalist – thinkers such as Hegel and
(even) Fichte.[11] They initially ran aground with liberal ministers in the Reich
Chancellery who were opposed to any state intervention with the economy[12].
Yet even though the liberal government generally repressed social democracy
plans, it could not circumvent the effects of the great economic downfall between 1873 and 1896. Faced with the rising voice by the *Arbeiterbewegung*
voting for the SPD[13], and the consequential social instability (partly due to the
Landflucht towards the urban centres of the Rhine-Ruhr region[14]), the government started to feel the need to offer workers and their financial dependants
more protection against the risk of industrial injury or death. Given the threat
coming from the vulnerable position of workers, the Iron Chancellor proclaimed an act which introduced compulsory insurance in 1881 in Germany,
which later ushered in modern social security laws.[15] As the legislator put it:[16]

---

[9] H. Barta (*supra* fn. 7), 86 claims employers' civil liability *vis-à-vis* injured workers yielded
'nur in sehr seltenen Fällen Aussicht auf Schadloshaltung'.

[10] K.A. Lerman, Bismarckian Germany, J. Breuilly (ed.), *Nineteenth Century Germany. Politics,
culture and society 1780–1918* (2001), 163–184 (on p. 175).

[11] In certain industries employers were already in the mid-fifties required to contribute to accident
funds administered by statutory associations. See J.D. Carr, Worker's Compensation Systems:
Purpose and Mandate, in: T.L. Guidotti/J.W.F. Cowell (eds.), Workers' Compensation (special
edition), [1998] Occupational Medicine 2, 417.

[12] Th. Nipperdey (*supra* fn. 6), 279 and 337 *et seq.*

[13] See for a captivating discussion H. Barta (*supra* fn. 7), 72 *et seq.*

[14] H. Barta (*supra* fn. 7), 57; V. Berghahn, Demographic growth, industrialization and social
change, in: J. Breuilly (*supra* fn. 10), 185–198 (on p. 188).

[15] H.G. de Gier/P.J. van Wijngaarden/A.M.E. Roelofs, *Sociale zekerheid in Europa: trends en
perspectieven* (1994), 36.

[16] *Verhandlungen des Reichstags,* 4th Legislaturperiode, IV. Session, III. Band, Anlage 2 zum
Aktenstück No. 41, p. 228.

'Dass der Staat sich in höherem Masse als bisher seiner hülfsbedürftigen Mitglieder annehme, ist nicht bloss eine Pflicht der Humanität und des Christentums, von welchem die staatlichen Einrichtungen durchdrungen sein sollen, sondern auch eine Aufgabe staatserhaltender Politik, welche das Ziel zu verfolgen hat, auch in der besitzlosen Klasse der Bevölkerung, welche zugleich die zahlreichsten und am wenigsten unterrichteten sind, die Anschauung zu pflegen, dass der Staat nicht blos eine nothwendige, sondern eine wohlthätige Einrichtung sei. Zu dem Ende müssen sie durch erkennbare direkte Vortheile, welche ihnen durch gesetzgeberische Massregeln zu Theil werden, dahin geführt werden, den Staat nicht als eine lediglich zum Schutz der besser situirten Klassen der Gesellschaft erfundene, sondern als eine auch ihren Bürfnissen und Interessen dienende Institution aufzufassen'.

8   The introduction of the industrial accidents insurance mainly came from the fact that there were only a few local health plans and liability law had not been rewarding. The main reason for the latter was that liability centred on fault, which generally needed to be proven by the victim-employee. The only exception to this fault liability was found in the *Prussian Railways* Act 1838, and this is generally explained by the fact that its regime of strict liability hardly affected the – mainly land-owning – dominant class because of the moderate territorial length of the railway tracks in those days.[17] The liability for railways that was introduced was intended by its instigator and Berlin professor Von Savigny to protect passengers and landowners (whose estates could be caught with fire by sparks from locomotives' smokestacks). The Prussian Supreme Court however reinterpreted this to include the protection of workers. Other workers were still left to fault liability.

Secondly, though reputedly as important[18], there was no uniformity of law. Germany better yet Prussia (*Preussen*) had as we know different jurisdictions (left aside the many *Partikularrechten* which were derived of each three): the Jus Commune (*Gemeines Recht*, the old common law of the Roman empire), *Allgemeines Preussisches Landrecht* (ALR, which applied in most of Prussia) and, covering a – from territorial as well as population point of view – minority part of the empire, *Rheinisch-Französisches Recht*. According to the first two systems' general civil law rules, one who by his negligence had caused harm to another could be held liable. This was not in the least pressing for accidents at work. The employer was usually absent from the workplace which made it practically impossible to prove fault for typically the injured worker himself, one of his co-workers or a third party other than the employer would be to blame. Also, the employer's risk of civil liability was particularly limited: The *Geschäftsführer (Unternehmer)* would only incur liability *vis-à-vis* his employees based on the theory of *culpa in eligendo* (and in the ALR only as a subsidiary rule of law). As we know today this latter theory actually rested on

---

[17] No more than 100 kilometres of tracks, according to B.S. Markesinis/H. Unberath, *The German Law of Torts. A Comparative Treatise* (4th edn. 2002), 724.
[18] H. Barta (*supra* fn. 7), 80.

a too extensive, too generalised interpretation by nineteenth century Pandec-tists of Roman law claiming that the employer could only risk liability if the harm had resulted from his own personal wrongdoing. In practice this meant that the employer would never be liable unless he too had been at fault, for in-stance by wrongfully selecting the employee who had caused harm to the in-jured co-worker or third party for instance (*cf.* currently § 831 BGB). On the contrary *rheinisch-französisches Recht* had been influenced by article 1384 of the French civil code and adopted a stricter rule, based on the principle of em-ployers' vicarious liability. As a result a *prima facie* liability was imposed on employers for negligent acts done by their employees towards co-workers or third parties (based on a refutable presumption of fault).

But after the unification of the *Reich*, in 1871, uniformity of law was enhanced by a new liability regime based on the so-called *Reichshaftpflichtgesetz (Lia-bility Act, RHpflG).*[19] It introduced a two-fold regime of strict liability. First, it imposed strict liability on the entrepreneur of the railways based on a '*prä-sumiertes Verschulden mit Beweislastumkehr*' (§ 1 of the Act). In addition, it exposed the entrepreneur of a mine, stone quarry, quarry or factory to the risk of vicarious liability (*erweiterte Haftung für Hilfen*, § 2 of the Act). Both grounds for liability were limited to damages caused by any man's injury or death. The railways regime did not require proof of fault by the victim-em-ployee (it did allow the employer to free himself either by an Act of God de-fence or by proving contributory negligence of the victim). The vicarious lia-bility rule for other industrial employers was less strict as it still required proof that the worker who caused the injury had acted negligently (similar to the *rheinisch-französischen Recht* discussed above), but this was made easier by a number of safety regulations that had been launched in the industry.[20] The em-ployer who had compensated his employee was given an action for reimburse-ment *vis-à-vis* third parties other than his employee based on the common lia-bility rules.

However as a remedy for the victim-employee vicarious liability actions were exceptional and hardly successful[21], as liability law had not yet attained its full development. In 1878 the liability regime of the Reichshaftpflicht-gesetz (*RHpflG*) was nevertheless extended to *operators* of gas pipelines, fumes and alike for damage caused to either *their* workers or third parties.

4.2.2 Strict Liability Subjected to Criticism: Arguments for a Shift in Governance

But even though the strict liability regime had become more protective, the system was unanimously found to be inefficient (and also, but less interest-

9

10

---

[19] 'Gesetz, betreffend die Verbindlichkeit zum Schadenersatz für die bei dem Betriebe von Eisen-bahnen, Bergwerke u.s.w. herbeigeführten Tödtungen und Körperverletzungen, vom 7. Juni 1871', *Reichs-Gesetzblatt* 1871, No. 25 (652).

[20] The new act thus probably had a moderate impact on the minority of Prussians to whom it applied.

[21] H. Barta (*supra* fn. 7), 102.

ing for our purposes, its protective ambit was received as being too narrow).
The difference between the two separate liability regimes of §§ 1 and 2 of
the act led to difficulties both in theory and practice.[22] On a theoretical level
there were doubts whether empirical findings would support and justify a
stricter liability regime for railway accidents compared to accidents in mines
or factories.[23] In practice complaints were made with regard to the heavy burden
of proof, as it was still partly related to fault.[24] The Act of God and contributory
negligence defences had remained essential barriers, while the vicarious liability
regime (§ 2 RHpflG) centred on the negligence of the employer's *Hilfe*. Private
insurers were free to use standard norms to deny or quantify damages, and the
risk of occupational disease was hardly protected. Injuries and disease were often explained to be the *'unvermeidliche Folgen der Beschäftigung in gewissen
Betrieben'* and this explanation was without further ado given a legal meaning
in the sense that the causal link between the accident event and the worker's
complaint was missing[25]. Overall the liability system was unsatisfying, or as
Bismarck himself would put it a good ten years later[26]:

> 'Es lässt sich hiernach nicht verkennen, dass der § 2 des Gesetzes vom 7.
> Juni 1871 der Absicht, den Arbeiter gegen die wirthschaftlichen Folgen
> der mit seinem Berufe verbundenen Gefahren sicher zu stellen, nur un-
> vollkommen entspricht, dass unter Umständen der Arbeitgeber durch die
> Haftpflicht in einer übermässigen Weise belastet wird, dass durch das
> Gesetz statt der gehofften Verbesserung des Verhältnisses zwischen Ar-
> beitgebern und Arbeitern in weitem Umfange der entgegengesetzte Er-
> folg herbeigeführt und im ganzen eine Situation geschaffen ist, deren Be-
> seitigung im Interesse beider Klassen der gewerblichen Bevölkerung
> gleich wünschenswerth erscheint'.

11   The amount of compensation which was offered by the liability system based
     on the *RHpflG* improved over the years (between 1874 and 1878 the awards of
     compensation were almost doubled). Clearly, this was to the expense of the in-
     dustry (as liability insurance premiums were multiplied by four).[27] Since the
     negative effects of the Act for the *victim*, as mentioned above, had also not
     been changed, *both* parties now heaped criticism on the system. Another com-
     plaint shared by both parties was that this – expensive – system had remained
     liability based so that it, as far as occupational injuries were concerned, jeop-

---

[22] H. Barta (*supra* fn. 7), 101: 'Als besonders gravierend empfand man die Haftungsdifferen-
   zierung der §§ 1 und 2 RHG, die den Ersatzanspruch völlig verschieden gestaltete'.
[23] H. Barta (*supra* fn. 7), 102 claims 'dass § 1 RHG die strengere Haftung für einen
   Wirtschaftssektor (Eisenbahn) wählt, der – schon bei einer bloss oberflächlich quantitativen
   Betrachtung – keinesfalls der drängendste war'.
[24] H. Barta (*supra* fn. 7), 102 quotes Riesenfeld complaining the 'Beibehaltung des Schuldmo-
   ments in § 2 RHG habe das meiste Unheil angerichtet'.
[25] See H. Barta (*supra* fn. 7), 134 *et seq.*
[26] Motiven zum Regierungsentwurf eines Gesetzes vom 8.3.1881 betreffend die Unfallversicherung der Arbeiter, *Verhandlungen des Reichstags*, 4th Legislaturperiode, IV. Session
   1881, 3. Band, Reichstagsdrucksache (RTD) No. 41, p. 230.
[27] Numbers taken from H. Barta (*supra* fn. 7), 98.

ardized the relationship between the employer and the injured employee and a good working environment (as far as there was any).[28] This risk of workplace conflicts, inherent to civil liability actions, was in fact the Prussian government's main argument for a retreat away from civil liability law and the subsequent shift to a compulsory no-fault insurance (which Bismarck had tried to enforce many years before):[29]

> 'Auf dem Wege der Haftpflicht werde das diesem Entwurf zugrundeliegende Prinzip des Schutzes der den Gefahren des Betriebes zum Opfer fallenden Arbeiter nur unvollkommen erreicht. Das Haftpflichtgesetz habe das Verhältnis zwischen den beiden Klassen der gewerblichen Bevölkerung eher verschlimmert als verbessert'.

Three more factors seem to have played an important role too. There were both '*dem zunehmenden Einfluss der Katheder-Sozialisten*'[30] and the Prussian government's fear that the protective and absolute nature of the industries' liability would soon be extended to other areas of liability law. Thirdly, Barta claims that in those years payments for an employee who had been disabled for two-thirds of his earning capacity, could easily amount to almost eight times the victim's net yearly income:[31]

> 'Ein Massenunglück hätte damit häufig den Ruin eines Unternehmers bedeutet und damit unter Umständen auch den Ersatzansprüchen der Verunglückten bzw. deren Hinterbliebenen die Grundlage entzogen'.

4.2.3 From Reichshaftpflichtgesetz to Social Reform: The Unfallversicherungsgesetz 1884 (Industrial Accidents Insurance Act, UV)

Not surprisingly therefore, once Bismarck's authority was growing and he had managed to be at the direct head of the Prussian Ministry of Trade, he got rid of the strict liability system and opted for the system he had long envisaged. In 1883 the *Gesetz betreffend die Krankenversicherung der Arbeiter* (Law concerning Health Insurance for Workers, KvA) was launched, the precursor of the *Krankenversicherungsgesetz* 1892 (Health Insurance Act, KvG). This introduced the first social medical insurance for most manual and white-collar workers in industry (although rudimentary health insurance programs had already existed at a municipal level). It was followed in 1889 by social security legislation which covered the risk of income loss due to old age and retirement. Based on the KvA the so-called *Krankenkassen* (national health services) offered prepaid medical care and also sick pay for the first thirteen weeks of disability.

---

[28] Th. Nipperdey (*supra* fn. 6), 341 reflects: 'Liebe und Furcht, so hat Theodor Lohmann gemeint, seien die Triebkräfte, die zusammen das Zustandekommen der Versicherungen ermöglicht hätten'.

[29] Motiven zum Regierungsentwurf eines Gesetzes vom 8.3.1881 betreffend die Unfallversicherung der Arbeiter, *Verhandlungen des Reichstags*, 4th Legislaturperiode, IV. Session 1881, 3. Band, Reichstagsdrucksache (RTD) No. 41, p. 232.

[30] H. Barta (*supra* fn. 7), 148.

[31] H. Barta (*supra* fn. 7), 142.

Shortly after this, and more importantly for our purposes, the Reichshaft-pflichtgesetz (Liability Act 1871, *RHpflG*) was abolished and in 1885 the so-called *Unfallversicherungsgesetz* (Industrial Accidents Insurance Act, UvG) of 6 July 1884 came into force. This special statutory Unfallversicherung (Industrial Accidents Insurance, UV) introduced a comprehensive system of public insurance,[32] which protected employees and their financial dependants against the risks of industrial accidents. The injured employee would be entitled to compensation for all personal damages such as medical health expenses, income loss with a maximum of two thirds of the victim's last salary; the first fourteen weeks would be covered by the general social security scheme as aforementioned. In the case of a workers death, funeral expenses were paid as well as a pension for life support of his widow and minor children. Fault was not relevant, except when the victim-employee himself had deliberately caused the accident. The statutory *Unfallversicherung* (Industrial Accidents Insurance, UV) was, and still is, financed by contributions of employers, administered by the so-called *Berufsgenossenschaften* (BGs) and supervised by the State (the *Reichsversicherungsamt*, which also functioned as a board of appeal). These *Berufsgenossenschaften* (BG's – as I will call them – are decentralized semi-public insurance boards of industrial branches, which represent both employers and employees but were supervised and controlled by the State. While the *Krankenkasse* awards as aforementioned were financed by both employers and their employees collectively, the *Berufsgenossenschaften* (BG) awards were paid by contributions of the employers only, based on accident rates in their branch.

14   It seems virtually impossible to exaggerate the impact of this shift from civil liability law to the new and compulsory, public law insurance system which, as will be seen, has remained in force. Not only did it work for insurers, as the market for life and health insurance gradually expanded in the years following 1884,[33] but also it seems to have been satisfying and effective for its purposes. Why? As was seen above, the initiative for this new insurance system primarily came from the desire to protect the working class against risks specifically associated with their activities. The statutory *Unfallversicherung* (Industrial Accidents Insurance, UV) was based on the so-called '*risque professionnel*': the costs of accidents – including the financial consequences of disability for workers – were taken to be part and parcel of production processes. The employer was found to be the better cost carrier, as he also gained from the production process and was in charge of it while directing his business entity.[34]

---

[32] In practice the administration of the insurance was highly decentralized.

[33] Th. Niperdey (*supra* fn. 6), 267 particularly points to life insurances: '... ihr Kapital war von 1880 bis 1913 von 0,44 auf 5,6 Milliarden Mark gewachsen ...; ähnlich war es mit anderen Versicherungen, Unfall-, Kranken- und Haftpflichtversicherung dehnten sich parallel zur Sozialversicherung stark aus'.

[34] For the same reason the *employees'* personal liability was, and still is, practically excluded in all situations where damage was caused to either his employer (business property), to his colleagues, or to third parties. Full liability with no means of reimbursement is regarded as being contrary to equity (*Billigkeit*), since the employee acts at the instance and in the interest of the

Yet undisputedly[35], employers benefited from the new insurance regime too, as it was a way to circumvent civil liability law and its growing expense. If employees were given a pension, social instability would not take a radical turn. The insurance model removed the uncertainty of negligence liability (and of the number of lawsuits brought by injured employees).

Of course this does not mean to say that the scope of protection offered by the insurance model was free from criticism. Some argued that the system created uncertainty as it was unsure what constituted an industrial accident (although this also had the advantage of making the system flexible; even in today's legislation there is no straight forward definition).[36] In 1925 for instance, as will be seen below, commuting accidents and certain other industrial damage events (some occupational diseases) would be included. Furthermore, the level of compensation of the new insurance scheme was moderate compared to the liability system it had replaced: as said, it provided for fixed medical aid and pensions were offered but only *after the fourteenth week of disability* (the first weeks being covered by mandatory health insurance for which both employers and employees paid contributions) and limited to, at the maximum, two-thirds of the employee's lost wages. Property damage and immaterial damages were not protected, which must have been to the relief of the industry (given the former increase of the premiums for liability insurance that had quadrupled in only six years under the – old – regime of employers' liability).

These limitations and the uncertainty of a sudden expanding scope of protection might well explain why the German compensation scheme was satisfying for the industry as well. Also for employees, it got rid of the uncertainty of law suits: with the exception of cases where they had intentionally caused the accident themselves, their right to compensation no longer depended on proof of how exactly the workplace accident had come about. What is more, as will be seen, a system of direct insurance such as this seemed more flexible in covering new accidents than liability law. Insurers would conform to the needs for recovery as they changed due to new industrial development and technique.[37]

In 1911 these mandatory work insurance plans as well as survivors' and orphans' benefits for work injuries were combined in the *Reichsversicherungsordnung* 1911 (National Social Insurance Act, RVO). As will be seen below,

---

employer. There has however been strong criticism on the matter, *e.g.* by B.S. Markesinis/H. Unberath (*supra* fn. 17), 706 *et seq.*, who claim that, legal technicalities set aside, the 'obvious favouring of the helpless employee may have much to support it. But it has still to be demonstrated how much it has added to the productions costs of German industry thereby making it internationally less competitive'. If this argument were to be taken seriously, the same point could be made for Dutch law which essentially alleviates the employees' liability in the same manner.

[35] Except for Bismarck's costly retirement and healthcare schemes, see V. Berghahn (*supra* fn. 20), 195.

[36] § 8 subs. 1 *Sozialgesetzbuch:* 'Arbeitsunfälle sind Unfälle von Versicherten infolge einer den Versicherungsschutz nach § 2, 3 oder 6 begründenden Tätigkeit (versicherte Tätigkeit). Unfälle sind zeitlich begrenzte, von außen auf den Körper einwirkende Ereignisse, die zu einem Gesundheitsschaden oder zum Tod führen'.

[37] B.S. Markesinis/H. Unberath (*supra* fn. 17), 726.

15

more recently this was replaced for the better part by the *Sozialgesetzbuch* (Social Security Code, SGB).

4.3 Liability Oriented Protection for Workers Compensation in England (1870–1920)

4.3.1 The English Employers' Liability Act 1880

16   The Unfallversicherungsgesetz (Industrial Accidents Insurance Act, UvG) is in contrast to the more employer-oriented position taken by the British Parliament regarding the workers' cause. Even though the process of industrialization had actually started here before spreading to Germany, it took a while before Parliament responded to its social disarray.[38]

There is no easy explanation for this. Hanes argues that labour unions in late nineteenth century England concerned themselves first of all with survival and securing higher wages instead of promoting compensation for workmen's injuries.[39] De Swaan explains how – compared to other European countries – elaborate regimes of poor relief took away much pressure. Next to Charity Organization Societies and local poor relief payments, workmen themselves had taken steps to minimize the financial risk of workplace injuries by contributing to employees' pools. More than in other countries these pools developed into serious compensation funds managed by autonomous organizations, the so-called *Friendly Societies*.[40] Their growth however only began to show in the 1880s, when the Liberal Government had already started to take action. Similar to German law, the cause of the working class was first addressed by enforcing a more strict liability regime which subsequently proved to be a failure. It was later replaced by social insurance.

In order to understand what brought about this drastic change from civil liability to public insurance in England, earlier steps taken must first be explained. What reasons did the liberals have for proposing a radical shift in governance from civil liability law to public insurance in the space of only four years? Was liability law not working for the same reasons as in German law?

4.3.2 Failure of the System of Civil Liability Law

17   Initially, in the late eighteenth and early nineteenth century, personal wrongs could only be remedied by a limited number of specific actions such as trespass on the case or nuisance.[41] In the following decades a general action in negligence gradually developed, but this hardly came to the injured employ-

---

[38] H. Barta (*supra* fn. 7), 65 (nt. 76a) notes regarding the explanation for this: 'Damit hängt wohl auch die 'exemplarische Bedeutung des englischen Arbeitsrechts für den Kontinent zusammen'.

[39] Later, in the twentieth century they would initiate social reform but not yet. See D.G. Hanes (*supra* fn. 7), 1 'the most that may be said is that they did not oppose it'.

[40] See, more extensively on this, D.G. Hanes (*supra* fn. 7), 75.

[41] B.S. Markesinis/S. Deakin/A. Johnston, *Markesinis and Deakin's Tort Law* (4th ed., 2003), 75 *et seq.*

ee's benefit as it required proof of fault *vis-à-vis* his employer. Besides, in the few cases where personal fault of the employer was shown, the latter would generally have a number of solid defences to his avail that would free him of liability.[42] It must be added that even long before the beginning of the nineteenth century employers not only risked liability for their own personal wrongs but also – vicariously – for the negligent acts of their employees, committed within the scope of employment. Different from the Prussian *ALR*, this did not require fault of the employer. This meant that even if the employer had carefully selected and supervised his employees he would still be liable for their wrongs.[43] The justification for this strict liability regime, based on the principle of vicarious liability, was under debate, but most agreed 'that the real reason for the employer's liability was that the damages were taken from a deep pocket instead of from the original tortfeasor, in all probability a man of straw'.[44] However, injured employees did not benefit from this principle of vicarious liability; based on the employer's defence of 'common employment' such an action would not lie if the injured employee and the employee who had caused his harm shared the same employer.

This remarkable position came from the English appreciation of the contract   18 of service, which was held to be an exclusive source of obligations imposed on the employer and the employee and would therefore rule out any damage claim in the absence of a breach of obligation in the course of employment. The principle of common employment did *not* affect industrial injury claims vis-à-vis the injured party's co-workers or the foreman, but both would probably not be insured. Hanes describes vividly how around the 1870s this position of common law, and particularly the theory of common employment, became in fact a political issue.[45] Employees who were informed about this or had experienced the denial of their claim because of common employment felt discriminated and in 1876 one of labour's principal spokesmen drew a bill in Parliament to abolish this principle altogether. To effect this withdrawal the Government appointed a committee of the House of Commons to inquire 'whether it may be expedient to render masters liable for injuries occasioned to their servants by the negligent acts of certificated managers [and alike, EE]'. As Hanes shows, in the same year the Royal Commission on Railways published its report, concluding that the law needed to be changed in order to

---

[42] The so-called 'unholy trinity' of defences, claims J.D. Carr (*supra* fn. 11), 419. This concerned the defences of common employment (now abolished), contributory negligence (fault by the victim which limits his claim for damages) and *volenti non fit injuria* (risk acceptance by the victim which bars a liability claim for injuries occurring from a known and obvious risk), see P. Cohen (*supra* fn. 7), 196, D.G. Hanes (*supra* fn. 7), 10 and, more extensively, P. Cane, *Atiyah's Accidents, Compensation and the Law* (1999), 273.

[43] For a discussion of the principle as applied *today* see B.S. Markesinis/S. Deakin/A. Johnston (*supra* fn. 41), 571 where it is added that this includes even those caused by frolics on their own. Even in the old days this principle of employers' vicarious liability included disobedient wrongs committed by their servants, see D.G. Hanes (*supra* fn. 7), 8.

[44] D.G. Hanes (*supra* fn. 7), 9.

[45] D.G. Hanes (*supra* fn. 7), 15 *et seq.* The facts as presented hereafter are largely taken from this discussion.

satisfy the railway workers' assertion 'that in numerous cases they are sacri-
ficed from causes and in circumstances which [but for the defence of common
employment, EE] would clearly give them a right to compensation'.[46]

19   Due to subsequent changes of Government and other – highly political – rea-
       sons it took until 1880 to finalize a bill to the effect that injured employees
       would have the same rights vis-à-vis their employer as any other third party
       stranger to the contract of service would have. But the consequences of the
       1880 Act from the employees' perspective were mainly negative.[47] The em-
       ployer's liability for industrial accidents remained fault based. More impor-
       tantly, the 1880 Act put a fixed limit on the amount of each claim, limiting it
       to the estimated earnings of the injured employee during the three years pre-
       ceding his injury. And facing the risk of more claims as a result of this Act,
       employers had started to contribute to *Friendly Societies* (which in effect
       could increase the employee's benefits by 25 percent), inducing their workers
       to in return, give up their right to legal action.[48]
           After this, new bills to improve the injured employee's position passed al-
       most every year thus making the issue one of considerable political weight.
       Both Labour and the Tories had now committed themselves to some modifica-
       tion of the law relating to employers' liability.[49]

       *4.3.3 Improving Civil Liability Law: The Employers' Liability Bill 1893.
               Goals and Purposes*

20   The first draft of a new proposed bill passed the House of Commons on 20
       February 1893. According to its explanatory memorandum as cited by
       Hanes[50], three arguments were put forward to change the law. Not only would
       the bill have to advance *equal treatment* of employees with regard to their in-
       ability of claiming damages from their employer compared to the non-work-
       ers' position (*e.g.* passengers were not barred in their recovery of damages for
       a train which had run off the line through the ignorance or want of skill of the
       engine-driver) but also it needed to discourage large businesses from escaping
       liability by delegating authority. Thirdly, enabling employees to claim damag-
       es from the employer would induce the latter 'to exercise that degree of care
       which his duty to his servants requires'.[51]

[46] The commission's words as cited by D.G. Hanes (*supra* fn. 7), 17.
[47] Cf. J.D. Carr (*supra* fn. 11), 417–422. D.G. Hanes (*supra* fn. 7), 21 strongly criticizes this so-
      called Employers' Liability Act 1880, as it curtailed the defence of common employment only
      in five specified sets of circumstances where the worker was injured by a negligent act of a
      foreman. The Act also limited the defence of Volenti non fit injuria in the sense that this would
      no longer be satisfied by the mere fact of being in service and is published as an Appendix in
      his work (*supra* fn. 7), 109 *et seq.*
[48] Surely this is not a negative as such but it *was* for the nineteenth century working class as their
      bargaining position was unequal to their employer's. See P. Cohen (*supra* fn. 7), 197 and D.G.
      Hanes (*supra* fn. 7), 23 *et seq.*
[49] Especially after the rise of the working class' votes due to the *Reform Bill* of 1884 which
      extended the franchise, see D.G. Hanes (*supra* fn. 7), 34.
[50] D.G. Hanes (*supra* fn. 7), 61.
[51] D.G. Hanes (*supra* fn. 7), 61.

Although for the sake of compensation a shift to an insurance regime would 21
probably have been more efficient, serious objections were put forward
against this. It was feared that a system of no-fault insurance like the German
one would not find sufficient public support and, on a more substantive level,
that it 'affords no security or incentive for the exercise of care on the part of
the employer'.[52] Introducing a new act, the *Employers Liability Bill 1893*, was
an attempt by the Government to legislate both compensation and accident
prevention. For that reason the new proposal included an injunction clause
which prohibited the continuing practice of enclosed agreements between employers and their employees to exclude civil law litigation. These so-called
contracting out agreements would secure employers that in return for their
contributions to the *Friendly Society* their employees would give up any action
for damages in the common law. Labour unions opposed this because of the
poor bargaining position of the victim-employee (who, based on these contracting out agreements, was 'forced' to give up his action for damages). So
too, the government feared that, similar as I see it to the argument made
against third-party insurance in more recent times, this would take away the
employers' incentive to take much needed precautionary measures to prevent
accidents. Or better yet:[53]

> 'The threat of litigation with all its attendant horrors of public exposure
> and possible cost tends to make the employer more careful and responsi-
> ble; whenever contracting out occurs, the employer, no longer threat-
> ened, turns to negligence and the incidence of accidents goes up'.

Accordingly the government proposed legislation that prohibited contracting 22
out clauses. Hanes argues that it did so to please the labour unions, whose
electoral support was needed.[54] But the House of Lords *rejected* the proposal
with its prohibition of contracting out; the employers that were represented in
the House feared being exposed to a new liability regime that they could not
escape by contracting out and a subsequent mass of litigation.[55] As a result the
legal position of injured employees and their financial dependants had not
changed. Where they had given up their action for damages they had to resort
to the limited awards of the Friendly Society or other funds; in other cases
they were forced to sue their employer (and to prove fault in order to be compensated). Clearly this adversarial nature of civil liability litigation is the main
reason that explains tort law's failure in those days, both in German *and* English law. Even if an absolute and 100 percent strict liability regime were imposed on employers, the individual worker's vulnerable and dependant position undermined the effectiveness of such a system as:[56]

---

[52] D.G. Hanes (*supra* fn. 7), 63.
[53] D.G. Hanes (*supra* fn. 7), 23 (argument taken from explanatory reports: Public Record Office
(PRO), Home Office (HO) 45/9865/B13816/123, 23) and 83.
[54] D.G. Hanes (*supra* fn. 7), 84.
[55] P. Cohen (*supra* fn. 7), 197 and see D.G. Hanes (*supra* fn. 7), 83.
[56] D.G. Hanes (*supra* fn. 7), 10.

'he had the odious task of proving that the man upon whom his future livelihood probably depended [or one of his co-workers, EE] had been negligent in the conduct of his business'.

4.3.4 From Fault Based to Strict – No-Fault – Liability: The Workmen's Compensation Act 1897

23 This pressing need to improve the position of employees in both ways (ex ante and by offering a more secure right to compensation) is what finally, in 1897, caused the entrenched middle-class in the House of Commons to make improvements.[57] The newly elected Conservative government, inspired by its two main leaders and advocates of social reform, Chamberlain and Salisbury, sought to address the workers cause too and in a successful way. The government saw two alternatives: either it could copy the old Conservatives' proposal of 1893, adding to it the power of contracting out, or propose a scheme of no-fault insurance.[58] For obvious reasons it opted for an intermediate position: strictly speaking still based on civil liability, though *de facto* the Workmen's Compensation Act 1897 introduced a form of public insurance.[59] Similar to German law this was presented as 'the most important and revolutionary piece of legislation in the nineteenth century'.[60] The fact that this new statutory regime received Royal assent so soon was due to the fact that it seemed beneficial to all parties concerned:[61]

> 'from the point of view of the worker, he received a legal right to compensation at no expense to himself; from the point of view of the employer, the cost of accidents was to be computed as another cost of production in the same way as the depreciation of capital assets [and calculated based on a fixed schedule, EE]; from the point of view of the common law, strict liability (liability without any imputation of fault) had been written into a statute'.

24 This new act imposed strict liability on employers, which entitled the injured employee and financial dependants in the case of death, to receive a limited amount of compensation for any accident arising out of and in the course of employment.[62] In practice this meant that the employer's liability came to be independent of the question whether or not there had been negligence on his part or of anyone employed by him.[63]

---

[57] See D.G. Hanes (*supra* fn. 7), 87.
[58] *Parliamentary Debates,* Fourth Series, 48, 1426, as cited by D.G. Hanes (*supra* fn. 7), 102.
[59] Chranston, R., *Legal Foundations of the Welfare State* (1991), Chapter 2 claims that the origins of the social security system go back to the Elizabethan age.
[60] Covering some 6,000,000 workers, see P. Cohen (*supra* fn. 7), 198.
[61] D.G. Hanes (*supra* fn. 7), 104.
[62] Similar to German law, the 1897 Act initially only applied to the railways industry, factories, mines, quarries, engineering works and certain buildings exceeding 30 feet high, chosen on the ground of being the most dangerous; see P. Cohen (*supra* fn. 7), 198.
[63] P. Cohen (*supra* fn. 7), 198.

The main conditions for holding the employer liable were now twofold. First, the accident needed to have arisen *out of* the employment, which meant that it had occurred as a result of the fact that the employee was doing what he was employed to do (as opposed to, for instance, getting hurt by lighting a cigarette).[64] The second condition for imposing liability on the employer was that the accident had arisen *in the course* of employment, meaning that the accident must have occurred during working hours. As in the German statutory (gesetzliche) *Unfallversicherung* (UV) this did include the time taken in getting to and from his place of work.[65] The employer could only free himself from this by showing that the accident had been caused by the employee's 'serious and wilful default' (the defences of *Volenti non fit injuria* and contributory negligence were thus limited to exceptional situations only). In 1906 even this defence was ruled out in cases of permanent disablement or death. In both cases a regime of absolute liability (*viz.* a guarantee that compensation would be owed by the employer) applied.

But an important difference was that, unlike German law, the English Act rendered the idea that the risk of industrial accidents was the responsibility of *both* employers *and* their employees. For this reason – set aside whether it was convincing – the awards in the case of a work related disability were limited by a statutory ceiling, initially to a mere fifty (!) percent of the employee's previous income. Similar limits were set on the compensation of the employee's financial dependants, in the case of fatal injuries. Lastly, the Workmen's Compensation Act 1897 set procedural rules incumbent on either or both the employer and his employees.[66]

All in all, the Workmen's Compensation Act 1897 introduced a strict, almost absolute liability regime that covered the risk of mainly income related damages. In this respect it strongly resembled the German system. The scope of protection was expanded over the years, first to agriculture in 1900, then in 1906 to all manual employees and to other employees provided their income was below a designated statutory ceiling and in that same year the expansion came to include certain occupational diseases.[67] Here too, resemblance with German law seems apparent. As in German law the statutory maxima on compensation awards were raised, first in 1906 and again in 1920. But stricto sensu the English system was part of liability law and was not regulated by mandatory public law. For one, the aforementioned practice of contracting out was still allowed for employees who participated in a scheme that was certified by

25

---

[64] 'Accident' is used in the popular and ordinary sense of the word as 'denoting an unintended and unexpected occurrence resulting in personal injury or death', P. Cohen (*supra* fn. 7), 213, though as will be seen above it soon came to include occupational disease as well.

[65] P. Cohen (*supra* fn. 7), 215. See for German law below, no. 63.

[66] For instance it prescribed that the employer had to keep an accident book of the particulars of the accident and that claims for compensation needed to be settled by either mutual agreement or arbitration, see P. Cohen (*supra* fn. 7), 234 *et seq.*

[67] Only after World War II, in 1918, the Workmen's Compensation (Silicosis) Act was passed, which offered a separate right to compensation, outside of the 1897 Act. See P. Cohen (*supra* fn. 7), 199 and more extensively on this J. Stapleton, *Disease and the Compensation Debate* (1986), 21 *et seq.*

the Friendly Societies. Furthermore, some employers were not liability insured, although most employers[68] either voluntarily *did* take out insurance or became members of mutual associations. *De facto* the 1897 Act thus perched third-party insurance, though:

- (a) not mandatory as in German law,
- (b) still based on the employer's personal liability, and
- (c) limited, as said above, to only fifty percent of the victim's damages.

Consequential to (b) was that it still forced workers to embark on costly adversarial procedures against their employer with the risk of losing. Consequentially to (a) it worked a bit less distributively than mandatory insurance would have since the latter, as is argued by Carr, more effectively spread its costs to the consuming public in the price of goods and services.[69]

4.3.5 From No-Fault Liability to Public Insurance?

26   Within a few years time however the liability system for employees was accompanied by *social security* insurance.[70] The National Insurance Act 1911, often referred to as the National Health Insurance Act, introduced social insurance for all kinds of health and welfare benefits, such as (free) medical aid and sickness benefits. It covered the risk of income loss due to illness or disability and unemployment of the whole industrial population.[71] In doing so, the state took on the responsibility to pay for basic personal detriment and the right to these benefits was made independent of the manner in which damages had been caused (*e.g.* whether due to a misfortunate fall on a skiing trip or a bad flue). Different from the employers' liability insurance, National Insurance had a mandatory public law character. It was financed through flat-rate contributions which came not only from employers and employees but also from subsidies by the state, raised through taxes. Benefits to replace *loss of income* however, whether due to sickness, disability or maternity leave, were – based on the National Insurance Act 1911 – *not* payable if the incapacity had been caused by an industrial accident or occupational disease, as compensation would already be secured by liability law under the Workmen's Compensation Act 1897.[72]

27   In 1925 this social security regime was expanded by the introduction of the Widows', Orphans' and Old Age Contributory Pensions Act, which included all persons who were insured under the National Health Insurance Act.[73] The

---

[68] P. Cohen (*supra* fn. 7), 200.

[69] J.D. Carr (*supra* fn. 11), 418.

[70] Plus the Industrial Assurance Acts of 1921 and 1923 which first regularised insurance companies and collecting societies selling funeral and life insurances and, by doing so, made industrial assurance an integral part of the national organisation of Social Insurance.

[71] *I.e.* persons engaged in any employment under a contract of service whose salaries did not exceed a statutory ceiling, including state officials, as defined by P. Cohen (*supra* fn. 7), 16.

[72] Unless the Workmen's Compensation would be less than the social benefit, in which case it would be deducted and the injured worker would be entitled to receive the difference from the National Health Fund, P. Cohen (*supra* fn. 7), 29.

[73] See on the contributory pensions acts more extensively P. Cohen (*supra* fn. 7), 59 *et seq.*

administration was state organized and put in the hands of *Friendly Societies* and private insurers, which were registered and approved by the government. As a result, all manual employees between the ages of 16 and 70 years old have mandatory insurance, as well as employed persons other than manual employees whose salaries were below a statutory ceiling.

Employees whose damages had 'arisen out of and in the course of employment' now had to rely on a combination of – internally different – compensation systems. For their medical treatment employees would have to resort to public insurance (based on the National Health Insurance Act 1911). Income loss was left to the Workmen's Compensation Scheme, that would compensate a limited amount of damages. Only in the – admittedly rather exceptional[74] – cases where the employer had not voluntarily taken out liability insurance based on this Act, could the employee resort to public insurance. These were important snags attached to the combination of both the Workmen's Compensation system and social insurance: Employees whose employer had not secured the risk of liability were treated differently from those who could in fact rely on the employers' liability insurance.[75] So too, mutual associations were a risk as there was no guarantee that sufficient reserves were being set aside each year to cover outstanding liabilities.[76]

28

Because of these complaints initiative was taken to change the law, but without result. A Departmental Committee was set up in 1919, to consider whether it would be desirable to shift the liability regime of the Workmen's Compensation system altogether into a system of accident insurance under the control or supervision of the State. That would have to replace the mixed regime for income loss (liability insurance or, if that had not been taken out, social insurance). The Committee basically supported this proposal by its finding that insurance, be it *liability* insurance, where it *had* been taken out worked quite satisfactory: It seemed profitable to the companies who sold it[77] and despite the quite excessive premiums they were charged, employers generally seemed satisfied. Payments by their insurance companies to the workmen seemed to be made promptly. Given the risk of financial inability where no insurance had been taken out and the excessive premiums where insurance was taken out, the Committee recommended mandatory liability insurance for employers *and* supervision by the State. It also proposed an increase of the awards for income losses, up to 66.6 percent of the average week income, while a voluntary agreement was made with insurers with regard to the pre-

29

---

[74] P. Cane (*supra* fn. 42), 274.
[75] If the employer *was* insured his rights against the insurer would transfer to and be vested in the injured worker, though at the same time the policy terms would bind the worker as they bound the employer. Where the employer would not be insured, the compensation was treated as among the classes of debts which needed to be paid prior to all other claims.
[76] P. Cohen (*supra* fn. 7), 201.
[77] P. Cohen (*supra* fn. 7), 200 claims it was 'a huge business – the premium income was £5,000,000, payments under policies approached £2,000,000, commission to agents accounted for £546,000, and management expenses were nearly £1,000,000'.

mium ratio.[78] But as will be seen below it would take another twenty years before action would be taken, albeit in a very different direction.

4.4 The Dutch Insurance Model for Industrial Accidents (1870–1920)

4.4.1 Workers' Insurance Replaces Fault Liability: Ongevallenwet 1901 (Industrial Accidents Act, OW)

30   In the same year the English no-fault liability Act of 1897 was proclaimed for employers, a similar legislative proposal was put forward before the Dutch Parliament but with no direct results. It would take a few more years before the Dutch Government would finally proclaim the Ongevallenwet 1901 (Industrial Accidents Act, OW), which was amended in 1921.[79] Prior to this, employees struggled with comparable problems as were discussed above. From the beginning of the nineteenth century they had been legally in a position to sue their employer for accidents due to the latter's negligence but in practice such claims only gradually started to set foot in the last quarter of the nineteenth century, from 1871 onwards.[80] In the Dutch transport, catering, medical and chemical industries for instance special funds were covering damages, which – after a modest start in the seventeenth and eighteenth century – slowly grew out to become professional insurances only covering those who paid contributions.[81]

31   The Ongevallenwet 1901 (Industrial Accidents Act, OW) introduced a mandatory insurance for employers with regard to the risk of work related injury or death suffered by their workers, which was made completely independent of civil liability. The risk of accidents caused by the vocation was perceived as a risk inherent to the production process. Therefore, its costs were part of the production costs that were to be borne by the one who also gained the profits: the employer.[82] This line of thinking seems exactly similar to German law (though different from the English – political – view of industrial accidents as being a shared responsibility of both the employer and employee). Not surprisingly therefore, the Dutch statutory insurance resembled greatly the German statutory (gesetzliche) Unfallversicherung 1884. It offered the injured employee or his financial dependents a direct right to compensation, funded by the employer.[83] The government made it its responsibility to ensure this le-

---

[78] The Committee made many more suggestions, for instance with regard to lump sum settlements, see P. Cohen (supra fn. 7), 202 et seq.

[79] Wettelijke verzekering van werklieden tegen de gevolgen van ongevallen in bepaalde bedrijven, Handelingen der Staten-Generaal, Bijlagen 1896–1897, 159. R.J.S. Schwitters (supra fn. 7), 24; effectively in operation in 1903.

[80] R.J.S. Schwitters (supra fn. 7), 11. From 1811 until 1838 fault liability was based on the French civil code and from then onwards on the – very similar – Dutch civil code, where in 1907 the more favourable regime of article 1638x (currently art. 7:658) was introduced.

[81] R.J.S. Schwitters (supra fn. 7), 13 et seq.

[82] P.S. Fluit/A.C.J.M. Wilthagen, Het risque social, in: A.Ph.C.M. Jaspers a.o. (eds.), 'De gemeenschap is aansprakelijk...' (2001), 107–125 (on p. 111).

[83] The explanatory memorandum of the Dutch OW 1901 abundantly shows that it was inspired by foreign systems of direct insurance. The German industrial accidents insurance is particularly

gal right to compensation, as liability law had left too many victim-employees uncompensated.[84] By doing so, it changed financial support for medical costs, work related disability and loss of life support from being a 'mark of favour' into *the legal right* to compensation.[85]

Opposite from the English Workmen's Compensation Scheme, the Dutch insurance scheme had a public law character. The obligation to pay compensation was *not* primarily a private individual's (*viz.* the employer's) responsibility but was shared collectively, by all employers, based on this statutory insurance. Employers would in a way gain from this new regime. Based on the Ongevallenwet 1901 (Industrial Accidents Act, OW) the injured employee's right to sue for damages was seriously limited, even beyond the insurance sum. The employer could in principle not be held liable for any damages of the victim-employee, certain exceptions left aside (see below). Not even for non-pecuniary damages, even though these were not covered by the accident insurance and had become recoverable in liability law after a famous decision of the *Hoge Raad* (Supreme Court, HR) in 1943.[86] This almost 'immunity' from civil liability was all the more important after a contractual prima facie liability was introduced in 1907, which was more favourable to the victim-employee. According to this rule the employer was liable based on a presumption of fault, the violation of a safety rule (article 1638x of the – former – Dutch Civil Code). Only a limited and small number of grounds could still give rise to this contractual or any tort based civil liability of the employer. The first ground for employers' liability to mention is where the income of the victim-employee would exceed a certain statutory ceiling; the employee could then still sue his employer for the remaining part of his damages. Secondly, liability law still applied for damages caused by wilfulness or a criminal offence committed by the employer. Thirdly employers were still exposed to civil liability law for any property damage that the victim-employee suffered as a result of his work. The exclusion of claims in all other circumstances was independent of whether the payments granted by the Ongevallenwet 1901 (Industrial Accidents Act, OW) would actually be received by the employee.

Employers were on the other hand now faced with the statutory obligation to pay for the entire insurance, in the form of insurance contributions (premiums). Certain employers could also offer their employees direct relief instead of taking out private insurance. Where the insurance had been taken out, the insurance contributions were calculated based on the accident risk of their branch: the higher the risk, the higher contributions would be (which can be taken to give employers good incentives to minimize this risk).[87] In practice

32

often referred to with regard to the scope of protection, see *e.g. Handelingen der Staten-Generaal*, Bijlagen 1896–1897, 159, no. 3 (MvT), 14–18.
[84] H.J.W. Pelster (*infra* fn. 182), 11.
[85] H.P.M. Adriaansens/A.C. Zijderveld, *Vrijwillig initiatief in de Verzorgingsstaat, cultuursociologische analyse van een beleidsprobleem* (1981), 17–19.
[86] HR 21 mei 1943, *NJ* 1943, 255 (Kreuningen/Bessem).
[87] B. Barensten, *Arbeidsongeschiktheid. Aansprakelijkheid, bescherming en compensatie* (2003), 83.

however, most companies that had their own medical service would handle claims themselves and pay directly to the victim, without reporting the accident to the public institution in charge of controlling and securing insurance protection, the *Rijksverzekeringsbank*.[88]

33   Similar to German law the '*risque professionnel*' was now *grosso modo* carried by employers, through insurance. Except for the aforementioned grounds for liability there was no personal or vicarious liability by the employer anymore; claims for damages were practically excluded. As said, an exception was made for the – rare[89] – situations of intentional wrongdoing or serious criminal offence by the employer and/or for property damage. Also employees whose income exceeded a certain statutory ceiling were, as was said earlier, still allowed a civil action for damages beyond the level of protection offered by the industrial accidents insurance. The argument which was put forward for limiting the employer's liability was found in the assumption that he should not have to bear the risk of the same costs twice. It was considered to only be fair to eliminate the employer's liability to the extent of the insurance that he himself had paid for by means of the insurance contributions. Furthermore, the replacement of the liability system and its adversarial nature, with the risk of high procedures, by this new insurance served to limit the risk of workplace conflicts[90].

34   Liability law also had a – small – role to play if the damage had been caused by third parties. In that case the *Rijksverzekeringsbank*, the State authority that granted benefits based on the Ongevallenwet 1901 and 1921 (Industrial Accidents Act, OW), was entitled to collect the OW pensions and benefits that were owed to the victim, back from the liable party. For those purposes it was given an action for reimbursement (art. 89 OW). Based on that action it would ask to be re-paid, while the victim – as far as he was entitled to these pensions and benefits – would lose his claim (since his damages would be compensated by the Ongevallenwet 1901 and 1921). In practice this meant that (proof of) the way the damage was caused was mainly still relevant in the few cases where liability law still operated: actions for reimbursement by the *Rijksverzekeringsbank* and the aforementioned cases in which the victim would still have a claim (see no. 33).

4.4.2 Level of Compensation; Revision of the OW in 1921

35   Similar to the German law, the Dutch *Ongevallenwet* 1901 and 1921 (Industrial Accidents Act, OW) offered compensation for accidents caused at work. This was taken rather strictly: Not just any injury suffered during working hours was taken to be a workplace accident, but only those which were effectively caused by the working activities part and parcel to the assigned job.[91] No

---

[88] B. Barentsen (*supra* fn. 87), 85.
[89] B. Barentsen (*supra* fn. 87), 12.
[90] H.J.W. Pelster (*infra* fn. 182), 19.
[91] B. Barentsen (*supra* fn. 87), 80.

compensation was made available for accidents wilfully caused by the injured employee himself or for income losses due to the incapacity to work lasting no more than the first few days. Both limitations served to prevent excessive claims.[92] The level of compensation was relatively high: The Ongevallenwet 1901 and 1921 (Industrial Accidents Act, OW) covered the risk of medical expenses, work related disabilities and loss of life support in the case of death. Income payments were limited, but still high compared to what the English liability regime of Workmen's Compensation (and even German law) offered: The injured employee would receive (at the maximum) 80 percent of his income for the first six weeks of his invalidity, with the exception of the first three days, followed up by pensions which were dependent of the seriousness of his injury and limited to 70 percent of the victim's average daily salary.

This latter restriction – the actual salary was disregarded, a fixed money tag was put on each specific handicap – led to much criticism as it left the victim-employee uncompensated if the monetary consequences outweighed the sum to be paid for his handicap. In this respect Dutch law was not much different from the other law systems under review.[93] Another objection was that the Ongevallenwet 1901 and 1921 (Industrial Accidents Act, OW) – similar to German law – only applied to certain, traditionally the dangerous branches of industry. However, as in German law, though less far reaching, it was extended to other trades (such as the Accidents at Sea Act 1919, the Agricultural Accidents Act 1922 and the Miners' Disability Act 1933).

4.4.3 Additional Protection. Social Insurance (Invalidity Act 1919)

Soon after this, a first step to a 'real' system of social security was taken, for the risks of old-age and 100 percent long-term or permanent disability. The so-called *Invaliditeitswet* 1919 (Invalidity Act, IW) offered the right to pensions for the risk of income loss of all employees whose income was below a statutory ceiling and who suffered from invalidity not caused at work. Based on social grounds – not on fault or the creation of risk – employers were expected to pay social premiums related to the employee's income. As these premiums, at least partly, were deducted from this income it was in fact the employee who paid for this arrangement. Klosse argues that this fit the principal rationale behind it, namely that the risk of old-age as well as the risk of at one point in life becoming handicapped were regarded as personal risks of the employee for which he was insured via his employer (and it would therefore only be just that he was effectively the one paying).[94] A similar statutory ruling was made to cover the risk of 100 percent *temporary* disability not caused by workplace injuries, which even offered sick leave based on at the maximum 80 percent of the last income, but this act only came into force in 1930.

---

[92] H.J.W. Pelster (*infra* fn. 182), 15.
[93] Cf. P.S. Fluit/A.C.J.M. Wilthagen (*supra* fn. 82), 110.
[94] S. Klosse (*infra* fn. 186), 94.

38   The result was that employees had social insurance (public law resources) for
      the risk of income loss due to handicap or old-age for accidents *not* related to
      work and 'private law accident insurance' protection for pecuniary damages
      due to injury or death which had been caused by workplace accidents. Only in
      the latter regime compensation was owed by the creator of the risk, the em-
      ployer.

4.5   Belgian No-Fault Liability and Voluntary Industrial Accidents Insurance (1870–1920)

*4.5.1   Arbeidsongevallenwet 1903 (Industrial Accidents Act, AoW) (I): Strict
          Liability*

39   The newly arisen state Belgium was the last to take action in the nineteenth
      century workers' cause, but there too social commotion forced the govern-
      ment into legislation. Initially employees were left to private initiatives or
      medical aid and – more exceptionally – the rules of liability law, that is: arti-
      cles 1382 (general clause of fault liability) and 1384 (vicarious liability with a
      reversed burden of proof) of the Belgian *Burgerlijk Wetboek* (Civil Code,
      BW), which forced them to take legal action against their employer and in
      principle to prove negligence (as article 1384 BW had not yet come to its full
      development from fault towards strict liability). Some good improvements
      were made to promote compensation. After the riots in the industrial cities of
      Liege and Charleroi in 1886, a semi-public institution was raised that had to
      encourage employers to take out insurance and that offered financial aid to the
      relief of employees who had suffered from an industrial accident.

40   Still it took the Belgian legislator until 1903 to introduce a basic right to com-
      pensation for employees and financial dependants who suffered damages from
      industrial injuries or death. The *Arbeidsongevallenwet* 1903 (Industrial Acci-
      dents Act, AoW)[95] introduced a strict liability regime for employers, inspired
      by the French act of 1898. It entitled employees who suffered from an indus-
      trial injury in the course of their employment to compensation and the require-
      ment of fault was removed altogether, except for where the injured employee
      had caused the harm by his own deliberate act. Arguably this regime of strict
      liability had a semi-public character: The Arbeidsongevallenwet 1903 repre-
      sented a deliberate move away from liability law's central focus on human
      acts and failure. Again, the sums to be awarded for industrial accidents were
      seen as part of the production costs: the damages that were caused in this way
      were, at least to some degree, considered to be an inherent part of the produc-
      tion process. Here too it was felt that employers had to bear these costs no less
      than they did with other production costs, regardless of fault. The Arbeidson-

---

[95] The Act of 24 December 1903 and, for occupational diseases, the Act of 24 July 1927 both
    were replaced by the Act of 10 April 1971 (Employment injuries) and the Laws co-ordinated
    by Royal Decree of 3 June 1970.

gevallenwet 1903 was in other words based on the 'risque professionnel', as in Germany and the Netherlands and therefore, similar to these neighbour law systems, it gave an almost absolute guarantee for compensation, independent of fault or human behaviour. Consequentially no Act of God defence or contributory negligence defence was allowed, as in the two aforementioned law systems, and the employer was free of additional *common law* liability (based on the general rules of article 1382 *et seq.* of the Belgian Civil Code).

In line with this view a mere causal link between the industrial accident and   41 the damages of the victim was needed for the victim's right to compensation, regardless of the way the accident had precisely come about. As in English, German and Dutch law the causal link was only accepted if the accident had occurred through and during work. If the latter was shown (an accident during working hours), it was presumed to have been a work related injury. Again similar to the German and Dutch systems of public insurance only intentional wrongs by the victim-employee could free the employer. The kinds of damages covered by the Arbeidsongevallenwet 1903 also showed great resemblance with German and Dutch law. The Belgian no-fault liability was limited to the victim's medical expenses and income loss and no awards were made for property and immaterial damages, for which the victim would also not have a general, common law action for damages.

But the level of compensation was, at least as far as income losses were concerned, much less far-going. Similar to the English system of strict liability, the victim received a mere fifty percent, at the maximum, of his *average* daily income in the case of full disability and likewise fifty percent in the case of a partial disability, though here: of his actual loss of income.

4.5.2 *AoW (II): Voluntary Insurance and the Algemene Spaar- en Lijfrente Kas (State Insurance Fund, ASLK)*

There was even stronger resemblance with English law, in the sense that the   42 obligation the Arbeidsongevallenwet imposed on employers was of a personal, private law nature. It was owed to the victim-employee, not to the state and there was no direct state control to see to it that the victim actually received compensation. For the sake of employees the Arbeidsongevallenwet *did* introduce a special fund where employers could insure against the risk of work related injuries or death in their business under the so-called *Algemene Spaar- en Lijfrente Kas*.[96] Similar to English law however, this industrial accidents insurance was made voluntary. The effects of this were less harsh by the introduction of the *Garantiefonds* (Guarantee Fund) which protected the employee-victim if his employer had not taken out insurance and was unable to pay compensation due to insolvency. But it would take until the 1960-s before the accidents insurance would be made compulsory.

---

[96] They could also take out insurance elsewhere. Fredericq, S., *Moderne risico's en vergoeding van letselschade*, (1990), 142; M. Faure/T. Hartlief (*supra* fn. 4), 94.

4.6   Comparison and a New Approach for All Four Systems

43   At the turn of the century the workers' cause was first addressed in Germany, then in England, the Netherlands and Belgium, by the introduction of schemes of industrial accidents insurance, exclusively financed by the employer. All four law systems share in common that they were expressly contemplated to replace the failing regime of fault liability by a more protective compensation system. These statutory occupational accident insurances provided benefits regardless of fault. Employers' and employees' statutory liability towards one another was transferred to their occupational accident insurance. Yet despite these similarities all four show great differences as to the end result. At the surface there is strong resemblance between German and Dutch early twentieth century law, both replacing big parts of liability law by no-fault insurance to be taken out and financed by employers for the benefit of their employees. So too, England and Belgium show great resemblance as both relied on a system of strict liability combined with voluntary liability insurance for extra protection. Although the latter solution might be explained as an insufficient response to the workers' cause, it also seems possible that due to the improvements of liability law in Belgium and England, there the need for a more radical approach was felt less pressing. Clearly the public law insurance systems of German and Dutch law were much more protective to employees as the right to compensation was made independent of the insurance contract concluded by the employer; it was ultimately the responsibility of the state.[97] All employees were insured by *law*, not by their employers' actions or consent.

44   Again let me briefly summarize the main features of the four law systems as were emphasized in the discussion above:

i.   First, the German *Unfallversicherungsgesetz* (Industrial Accidents Act, UV) and the Dutch *Ongevallenwet* (Industrial Accidents Act, OW) had replaced the 'normal' liability regime for employers by a mandatory no-fault insurance for employers, while in Belgian and English law employers could be held to a rather strict personal obligation based on personal liability.

ii.   In Belgian and English law, strict – almost absolute – liability regimes offered more protection for the 'happy few' as liability was made irrespective of the employer's behaviour, while on the large scale basic insurance protection was missing. In both German and Dutch law the liability of employers remained to be fault based with a reversed burden of proof. My guess would be that the direct reasons for this were political in nature: There seemed little room for a new and more protective, strict liability regime as employers already needed to carry the risk of insurance. But surely too it had to do with the fact that the new insurance had introduced a relatively high level of protection so that the need for a more strict liability law was felt less pressing. The remaining risk of employers' civil lia-

---

[97]   H.J.W. Pelster (*infra* fn. 182), 28.

bility was limited to exceptional situations such as intentional wrongdoing by the employer and for minor kinds of damages.

iii. This also seems to have affected the level of protection, albeit not in the way that would be expected. German and Dutch law – based on insurance – had a more narrowly defined scope of protection: both insurance systems only applied for personal damages and there were fixed awards. In both Belgian and English law however, the level of compensation (the amount of damages) was more modest: In the case of disability the victim would receive no more than fifty percent of his loss of income. I suspect this comes from the fact that the obligation to pay these awards primarily rested on the individual employers shoulders and was not borne collectively.

iv. There were differences between German and Dutch law as regards the social context in which the insurance came about. Prior and next to the employers' accident insurance, German law had a refined system of social security paid by the contributions of employers and employees, which covered a small part of the damages too, *viz.* the first weeks of income loss. Dutch law was at this time, before 1920, not as far; it had social security for the risks of old-age or long-term disability, paid by the contributions of employers and employees, but both social security schemes applied to causes of damage other than industrial accidents (which were ruled exclusively by the Ongevallenwet 1901 and 1921).

v. In England and Belgium third-party liability insurance was available, but this was not mandatory and was – as for England – still lacking any form of statutory ruling/protection. In German and Dutch law on the other hand the insurance was mandatory for employers as well as the minimum level of insurance coverage.

vi. All four systems forced the legislator repeatedly to (re)take action as the need for protection became more compelling. The German and Dutch system made direct state intervention necessary, as a system of mandatory insurance can only work if employers keep their obligation to insure. This forced the Dutch and German apparatus to control and supervise employers and the private insurance companies, resulting in the maintenance of a special Civil Service to perform these public law tasks. To some authors this made the workplace insurance overtly expensive and complicated, compared to payments coming directly from the State, as will further be seen below.[98]

5. Shift II (1920–1970): From Private to Social Insurance

5.1 The Need for More Protection: 'Risque Social' Thinking

Throughout the first half of the twentieth century the need for collective protection through public law intervention gradually grew stronger. The promot- 45

---

[98] H.J.W. Pelster (*infra* fn. 182), 27: 'Dit is een ongewenschte toestand van halfslachtigheid. … De wetgever [heeft te zorgen], dat er een of meer staatsverzekeringsinstellingen zullen worden gesticht'.

ing of stable economies after World War I had been the starting signal in most European countries for revisions of private accident insurance regulations and the introduction of public law insurance administrated by the state and financed by public law premiums or taxes. The post-World War II economies set the growing concern of social welfare in motion. De Gier, Wijngaarden and Roelofs explain how poor living conditions in the cities, due to the continuing process of urbanization, and the transition from agrarian to industrial economies, both created a growing need for protection from the risk of income loss and medical expenses.[99] They also point to factors such as the Enlightenment belief that poor social conditions were the result of human action and could be changed, the introduction of voting rights and the formation of Central Statistical Offices showing clear numbers and information on social conditions.[100] All of this added up to a growing need for social security.

46   The basic idea was to offer citizens protection for certain risks threatening their means of living (illness, work related disability, unemployment, old age and the financial dependants' risk of death). The Atlantic Charter had laid down certain basic rights commonly shared. Social insurance schemes were introduced which laid down the obligation for all insured persons and even uninsured persons to participate, in order to make new public compensation schemes payable, without risk differentiation. Low risk persons (young and in good health, etcetera) did not enjoy preferential treatment or reduction of their premiums.

5.2   English Shift to Its Current Regime of Compensation (1920–1970)

5.2.1 Beveridge's Plea for More Protection

47   In the post war years employees suffering from workplace accidents in England were protected by strict liability with voluntary liability insurance based on the Workmen's Compensation Act 1897 (WCA) for their loss of income. As for their medical expenses and some other injury related damages the system was, completed by the National Insurance Act 1911 scheme of social insurance. But more and more this system heaped criticism.[101] From the 1920-s the main issue had been that the level of compensation of the Workmen's Compensation Act 1897 (WCA) was too low and that given the voluntary nature of liability insurance, this resulted in unequal treatment of employees. Through the years this had gained force by the support of trade unions, probably mainly due to the great depression[102], followed by persistent inflation after the Second World War with the increase of bankruptcy and insolvency. The in-

---

[99] H.G. de Gier/P.J. van Wijngaarden/A.M.E. Roelofs (*supra* fn. 15), 38.
[100] See, more extensively, H.G. de Gier/P.J. van Wijngaarden/A.M.E. Roelofs (*supra* fn. 15), 39.
[101] See *infra*, para. 4.3.5.
[102] P. Cohen (*supra* fn. 7), 14 notes that in the period 1921–7 alone, claims for social sickness benefit of men had arisen by 41 percent, of unmarried women by 60 percent and of married women even by 106 percent. Even more so, corresponding social disability benefits showed increases of 85, 98 and 159 percent.

flation considerably reduced the value of workers' compensation liability insurance *and* national insurance, both of which were related to contributions made years earlier.[103]

All of this forced the state to re-evaluate both compensation schemes, which led to the famous Beveridge Report, in 1942. Beveridge wanted the government to secure a shared minimum level of existence, equal for all citizens, funded by flat-rated and equally high premiums for all. The centrepiece was a state-run system of mandatory insurance. By contributing to a scheme of national insurance deducted through the weekly or monthly pay packet, every employee, even working married women, would be helping to build up a fund that would pay out weekly benefits to those who were sick or unemployed or who suffered from industrial injuries. Hence, the Beveridge report put forward a serious critique on the *Workmen's Compensation Act 1897* (WCA). It argued that this scheme relied on individual personal liability and expensive private insurance. As for the latter, the Workmen's Compensation Act 1897 (WCA) offered a compensation scheme which was large in the number of persons that it covered though not in proportion to the annual sum of compensation that was paid. Furthermore, the report rejected the adversarial nature of the Workmen's Compensation Act 1897 (WCA), which forced unequal parties to settle by bargaining and the fact that it permitted payment of, as the report reads, 'socially wasteful lump sums' instead of periodical pensions in cases of serious incapacity.[104] As a result Beveridge in principle favoured to integrate the Workmen's Compensation Act 1897 (WCA) and the envisaged National Health Insurance. Much was to be said, he argued, to then make the former subject to the limitations of the latter, since:[105]

> 'it might well be argued that the general principle of a flat rate of compensation for interruption of earnings adopted for all other forms of interruption, should be applied also (…) to the results of industrial accident and disease (…). If a workman loses his leg in an accident, his needs are the same whether the accident occurred in a factory or in the street (…) Adoption of a flat rate of compensation for disability, however caused, *would avoid the anomaly of treating equal needs differently* (…)'

On certain aspects however, the report still held arguments to give employees suffering from industrial injuries a separate and more favourable position than other employees or others within the national system, for instance by offering earnings (not flat) rated pensions for *long-term* disablement caused by work related injuries. This would be fair for at least three reasons.[106] Firstly, employees must follow orders which make them subject to dangerous operations (and are therefore more vulnerable than victims who are left 'free' in their whereabouts). Also, more compensation would serve as an incentive to work-

48

---

[103] P. Cane (*supra* fn. 42), 276.
[104] 'Beveridge Report', *Social Insurance and Allied Services*, Report by Sir William Beveridge, Command Paper 6404, London: Her Majesty's Stationary Office, 1942, 80.
[105] Beveridge Report (*supra*, fn. 104), 80 Italics added, EE.
[106] Beveridge Report (*supra*, fn. 104), 81.

ers to accept hazardous jobs. And lastly, more protection through the Workmen's Compensation Act 1897 (WCA) scheme would enable the government to limit employers' liability at common law, which was found necessary given the aforementioned complaints with regard to its adversarial nature. Particularly this third argument seems interesting. It shows that one of the main reasons behind the – as will be seen: current – insurance system is that liability law was rejected, reputedly for its adversarial nature but perhaps more so for the financial burdens it laid on employers (enterprises) as it was partly replaced by a mainly tax funded compensation system, see below.

5.2.2 More Protection I: From Liability Law to Social Scheme, the National (Industrial Injuries) Insurance Act 1946

49 The British government in a sense shared the report's passage as cited above;[107] it abolished the Workmen's Compensation Act 1897 (WCA) scheme altogether and replaced it by two renewed schemes of social insurance.[108] First it introduced a renewed National Insurance Act 1946, following the prior act of 1911 which as said had been completed by benefits for (wrongful) death and other, for our purposes less relevant risks such as old age, retirement and maternity, suffered by all citizens, regardless of the manner in which the damage had been caused. Here, employees were thus not in any way given more protection than others with sickness or disability were. Secondly though, a special – more favourable – regime of social security was introduced which covered the risk of *income loss* and offered special treatment for workplace victims. The latter so-called National Insurance (Industrial Injuries) Act 1946, which came into effect on 4 July 1948, offered higher rate disability benefits for employees who suffered from workplace accidents or occupational diseases. The introduction of this scheme implicated a substantial shift in at least four ways:

(a) Social security now came to protect both the risks of medical expenses and income loss for victims of workplace accidents and occupational diseases.
(b) The risk of income loss was still governed by a special regime, but this was part of social security and was undoubtedly more favourable to employees who were disabled as a result of workplace accidents or occupational diseases. They would receive more and higher rated pensions.
(c) All of these benefits and payments, including the disability payments, were now financed by employers, employees and the Exchequer (mainly the tax payer).
(d) It was now the State, not the employer or his insurance company, who carried responsibility for all of these damages.

In the following years the social security scheme was extended even further. In 1948 the National Assistance Act was passed which offered benefits for those not covered by the National Insurance Act. National Assistance Boards were

---

[107] Government White Paper, *Social Insurance*, Part II, Command Paper Her Majesty's Stationary Office 6551, 1944, 26.
[108] It also launched, for our purposes less relevant, plans which completed social insurance even further, see P. Cane (*supra* fn. 42), 276.

set up to help citizens whose resources were insufficient to meet their needs. However, benefits were set too low which resulted in many citizens remaining below the subsistence level.

The benefits offered by the preferential regime for workplace victims were   50
still relatively limited however; the system did not offer full compensation. Beveridge's plea for a *preferential* regime for workers within the national system was *not* fully supported by the government, but in a more limited way it did favour victims of workplace accidents: They would already be entitled to higher rate disability pensions from the first week of disability onwards. But their *actual* loss of income was *not* fully compensated. Their pensions were made equal to those of employees whose inability to work was *not* caused by work related injuries. Compensation of income losses would therefore continue to depend on the degree of incapacity and *not* on the actual income loss. Clearly, since the level of compensation of this new social scheme was moderate in general and particularly limited for income loss, and as the strict liability Act of 1897 had been abolished, extra protection in the common law was still in order. Subsequently liability law was not completely set aside: The employee would be allowed to seek additional protection by bringing in a common law action for damages *vis-à-vis* the employer (*i.e.* liability based on the tort of negligence or other torts).

5.2.3 More Protection II: Employers' Liability (Compulsory Insurance) Act 1969 (ELA 1969)

I would guess partly influenced or led by this disappointing level of compen-   51
sation and despite the prior objections as mentioned above, liability law too gradually improved and from the 1940-s onwards one could say it had truly started to fill the gaps left by social security. The legal position of employees who claimed damages for workplace accidents in liability law grew stronger in at least *two* ways. First there was common law liability. As was said earlier, throughout the nineteenth and early twentieth century a common law action would rarely be worthwhile. One of the reasons for this was that common law liability of the employer *vis-à-vis* his employees was still based on personal fault, which could hardly be shown. Employees did not benefit from the employer's *vicarious liability* for wrongs committed by co-workers, because the employer could raise the defense of 'common employment'. Over the years this defence, based on an imaginary contract (*i.e.* the contract of service as interpreted by the courts), had been the subject of much criticism. As a result it became a dead letter in case law, in the 1930-s and finally it was abolished altogether by the Labour Party's Law Reform (Personal Injuries) Act 1948. The second improvement of the legal position of employees suffering from workplace accidents came from special legislation imposing specific duties of care on the employer, particularly the Employers' Liability (Defective Equipment) Act 1969). If it could be shown the employer had breached any of these duties, the victim-employee had an action for damages based on liability law (the tort named breach of statutory duty).

52   More protection was given by the Employers' Liability (Compulsory Insurance) Act 1969 (ELA 1969). This new act made liability insurance mandatory for all employers except for nationalised industries, the police and local authorities.[109] The failure to insure came to be listed as a criminal offence. The Employers' Liability (Compulsory Insurance) Act 1969 (ELA 1969), which is still in force today, introduced a system of insurance compensation. The insurance protection scheme strongly resembled German law, but there the employers' personal liability was no longer the legal ground for protection. A similar approach was considered in England as well, in the early 1970-s, but it was feared that insurance *not* based on liability (and liability law's principle of full compensation) would bring about a lower level of compensation.[110] Therefore the workers' compensation scheme remained to rest primarily on social insurance and in addition came an improved system of fault liability (either vicarious or for the employer's own breach of statutory duty).[111] Similar to Dutch law, employers' liability gradually expanded, for instance to 'safeguarding his employees against certain kinds of psychiatric harm'.[112] The rationale of employers' liability was found in the view that the employer was the one in control of and taking profit from the work being done, plus he was found to be the better risk carrier. The threat of personal liability which was now more real than ever would become an actual deterrent, next to special statutory safety duties imposed on employers and the obligation to report accidents and accident risks.[113]

5.3   Dutch Shift to Its Current Regime of Compensation (1920–1970)

5.3.1   Social Insurance: From 'Risque Professionnel' to 'Risque Social'

53   A similar development, though much later in time took place in the Netherlands, although the starting-point was completely different from England. As we have seen, Dutch law, once faced with industrial accidents, *had* opted for a statutory, mandatory no-fault insurance based on the *Ongevallenwet* (Industrial Injuries Act, OW). As employers were already carrying the cost of industrial accidents this way collectively, Dutch law – opposite to the British 1897 Act – strongly *limited* the employer's risk of civil law liability.

The scope of protection partly improved through the following years, particularly in 1921. In that year industrial accidents insurance was made compulsory for *all* businesses and branches, dangerous and non dangerous and in principle regardless of size. Secondly it came to cover any injury or death related to work. Thereafter it also came to include coverage for the risk of certain occupational diseases, listed in the revised Act, and accidents on the way to and from work.[114]

---

[109]   B.S. Markesinis/S. Deakin/A. Johnston (*supra* fn. 41), 561.
[110]   B.S. Markesinis/S. Deakin/A. Johnston (*supra* fn. 41), 561.
[111]   Or vicarious liability; B.S. Markesinis/S. Deakin/A. Johnston (*supra* fn. 41), 561.
[112]   B.S. Markesinis/S. Deakin/A. Johnston (*supra* fn. 41), 563.
[113]   In the 1990s the Health and Safety Commission estimated that about a third of reportable nonfatal injuries are in fact reported, see P. Cane (*supra* fn. 42), 177 (nt. 10).
[114]   P.S. Fluit/A.C.J.M. Wilthagen (*supra* fn. 82), 110.

But similar to England the level of protection for employees was perceived as   54
being insufficient as the concept of social security was starting to take form,
after World War II. In 1945 the so-called Van Rhijn committee was formed by
the Minister of Justice, to examine the options for forming a scheme of social
insurance. This committee felt that public insurance would have to be carried
by the public at large; to put it bluntly, the more collective the paying community the more certain the insurance would be. For this reason it was doubted
whether the 'risque professionnel' would still serve as the foundation for public insurance and the committee also pointed at the benefits of not having to
decide when and if the cause of the accident was in fact work.[115] Nevertheless
it decided *not* to extend the 'risque professionnel' protection of the Ongevallenwet 1901 and 1921 (OW) to all injuries regardless of their cause. It feared
this would cost too much and also hesitated to deprive employees of their special treatment. Towards the end of the fifties however, the more favourable position of victims whose injuries were caused at work, received more criticism.
The idea started to set foot that the state should guarantee a minimum living
standard for all its citizens in all situations, as general as possible, based on
the so-called 'risque social'. Moreover, this welfare state should intervene
with the process of economic reproduction and distribution to reallocate life
chances of individuals or classes of individuals regardless of their cause.
Again it was argued that the Ongevallenwet 1901 and 1921 (OW) had been
based on the 'risque professionnel' theory, which only aimed at protecting the
working class against risks which were typically associated with hazardous
employment activities.[116] Inspired by Beveridge, the idea now set foot that a
minimum standard of living should be guaranteed in the case of sickness or injury, and that those risks would then best be spread by the public at large.

Even more than in England, the Dutch social security system was fully found-   55
ed on this principal idea and it rejected any distinction between damages
caused at work and other causes altogether. Different from England it refused
to give employees who were injured at work any form of preferential treatment. The basic protection was given to both citizens and employees – regardless of how their damages had been caused (*i.e.* in or outside of employment).
As a result the Ongevallenwet 1901 and 1921 (OW) was taken out of force in
1967 and was – roughly put – replaced by the current regime of social security. Interestingly, the separate and more favourable compensation system for
state officials, financed directly by the state, was *not* abolished and thus continues to exist separately and offers greater protection in the case of injuries or
death caused at work. As said the officials' position will further be left aside
here.

---

[115] P.S. Fluit/A.C.J.M. Wilthagen (*supra* fn. 82), 115.
[116] W.J.P.M. Fase, De legitimering van het verplichtend karakter van de sociale verzekering, in:
A.Ph.C.M. Jaspers (*supra* fn. 82), 52.

5.3.2 Three Pillars of Social Security

56   This newly formed social security system came to rest on three pillars: the national insurance regime, the employee insurance regime and the regime for self-employed persons. It was mainly paid for by both employers and employees through contributions independent of the nature and seriousness of the risks that were covered and through taxes.

57   The first pillar found its roots in the late fifties and sixties. National insurance was introduced, to serve all residents in the Netherlands who were not excluded from coverage under national or international regulations. Influenced by the Beveridge model this scheme of social security was intended to offer basic protection to all citizens and was partly set up as national insurances (*volksverzekeringen*) and partly as special social services (*sociale voorzieningen*). It applied to both workers and non workers. Based on the idea of a welfare state a number of regulations were put forward, introducing flat rate schemes of social insurance or funds such as insurance for financial dependants in the case of death, child care, a general health insurance scheme, a disability insurance scheme, retirement pensions and unemployment benefit schemes. Insurance schemes of this kind were put into force by the (former) Algemene Weduwenen Wezenwet (Surviving Dependants Act, AWW) 1957 – (this Act would later be replaced by the Algemene Nabestaandenwet), *Algemene Wet Bijzondere Ziektekosten* (Exceptional Healthcare Act, AWBZ) in 1968 and by the *Algemene Ouderdomswet* (Old Pension Act, AOW) 1957 and the *Algemene Kinderbijslagwet* (Child Care Act, AKW).[117] The last two Acts will not be of much interest here, as both will hardly ever coincide with civil liability law since they cover the financial risk of old age and life support expenses for raising children. The national insurance laws of this regime that do seem relevant for our purposes share in common that they were mainly introduced to cover special long term medical expenses and pensions for financial dependants in the case of death. Later, in 1975 the (former) *Algemene Arbeidsongeschiktheidswet* (General Disability Act, AAW) was put in force which was most significant as it first came to cover the risk of income loss in the case of illness or disability of all Dutch citizens. More recently, in 1998, this Act was replaced by more specific legislation (*viz.* WAJONG).

58   The second pillar of the social security program that was developed was a special regime for employees, the 'employees' social insurance' (*werknemersverzekeringen*). This was set up after Bismarck's insurance *Krankenversicherung*, offering special protection for injured employees and their financial dependants regardless of whether the injuries or disease had been caused by workplace accidents. It was actually introduced in 1967, as the Ongevallenwet 1901 and 1921 (OW) was then replaced by social health and disability in-

---

[117] The *Algemene Ouderdomswet* (General Old Age Pensions Act, AOW), which replaced the old Invalidity Act of 1919, provides insurance coverage for the financial consequences of having reached the age of sixtyfive. This retirement scheme may further be left aside here as it can hardly be concurrent to tort law's scope of protection.

surance schemes. This meant a dramatic change as new, social security systems now came to replace the 'risque professionnel' coverage (protection for the risk of industrial injuries) by offering personal injury compensation for any working citizen that suffered damages – regardless of the way these damages were caused. They no longer depended on the employers' obligation to take out insurance; the victim was entitled to benefits by an act of law. This special part of social security law includes several social insurance schemes which offer protection against the risk of loss of income and exceptional expenditure due to old age, death, illness, disablement or unemployment. The most important scheme of all[118] was the introduction of the (former) *Wet op de arbeidsongeschiktheidsverzekering* (Disability Insurance Act, WAO) in 1967, for the risk of long term income loss. For the first short-term period of disability, the Ziektewet (*Sickness Act*) entitled the employee to sick pay (and also maternity leave). Based on the former WAO, employees under the retirement age of sixty-five would thereafter be given the right to benefits, that is: if they had been disabled for more than fifty-two weeks. More recently, in 1998, this regime was partly replaced by another scheme for long term disability, see nos. 81 and 82 below. The risk of medical expenses came to be protected by the (former) *Ziekenfondswet* (National Healthfund Act, ZFW), covering salaried persons earning less than a certain amount of wages. Since the beginning of last year, 2006, this latter Act too has been abolished and replaced by a general statutory insurance for *all* citizens, based on private insurance, resulting in a clear cut (external) shift in governance: The risk of basic medical costs of employees – either work related or not – is still carried collectively, but it is now based on private insurance instead of social insurance. 'Special' expenses such as highly expensive and/or long lasting costs such as hospital bills or the costs of reintegration facilities were, and still are, covered by the *Algemene Wet Bijzondere Ziektekosten* (Exceptional Healthcare Act, AWBZ). The so-called National Healthfund thus came to provide coverage for all the (para)medical costs of health care, including hospital, revalidation, sanatorium and maternity care costs. It was made compulsory for employees whose earnings did not exceed a specified threshold; others had to, and still must, take out private health insurance. Non pecuniary damages were left to civil liability law.

The third pillar of social security is less relevant for our purposes as it does not cover injury risks directly and it will hardly ever coincide with liability law. It contained, as it still does, social welfare and unemployment schemes and alike. It served as a rest category, and it still does. The *Algemene Bijstandswet* (Social Welfare Act, ABW) seeks to protect the long-term unemployed and certain other categories of individuals against the risk of income loss, provided they have no other resort such as the long-term unemployed who are no longer entitled to income related payments and persons who are not in salaried

59

---

[118] Left aside that in addition to the OW which covered the 'risque professionnel', the *Invaliditeitswet* (Invalidity Act, IW) had been introduced in 1919, covering the risk of invalidity caused in any other way but through work ('risque social').

employment, such as self-employed business owners or freelancers. Public officials were entitled to extra benefits, covering all their (additional) medical expenses and income losses. For that purpose a special statutory regime was introduced, which – interestingly – offered even more – 'super' – protection (*i.e.* higher awards and alike) if their medical expenses and/or income loss had been caused by work related injuries.

More recently, in 1998, public officials were integrated in the 'regular' schemes of social security, which means that their income losses and medical expenses are basically covered by the same funds and legislation as regular employees' damages. But public officials still enjoy additional protection in the sense of higher awards and longer duration of payments. Given the great variety of arrangements in this respect, the position of public officials will further be left aside and the focus will be on the rights to compensation of regular employees.

60   Next to the social security schemes discussed above, replacing the Ongevallenwet 1901 and 1921 (OW), the risk of unlimited civil liability was now reintroduced. This meant that in *all* cases (read: also in cases of mere negligence) the employee could bring a civil action against his employer for damages *that were not already covered by any scheme of social insurance.* For the part of the damages that were covered by social insurance, the social insurance carrier was given an action for reimbursement against the liable person (the employer, a colleague of the victim or in principle any third party who caused the accident). If directed at the employer or a fellow-employee of the victim-employee however, these actions were, and still are, limited to cases where there was gross negligence or intentional wrongdoing by the employer or the fellow employee.

Not surprisingly therefore, there were not too many actions in the first half of the twentieth century. On the part of the employee one obvious reason for this is that employees feared to disturb the working relationship as it was – some fifty years ago, that is – not common to sue their employer. Furthermore, it seems likely that there was less need for a civil law action than there is today as social insurance coverage was more satisfying. It was assumed unusual for employees to sue and there were fears that this would jeopardize their position at work. Moreover, the costs and energy of a civil law action for the relatively small amount of damages that were left uninsured probably did not seem rewarding. Social security carriers were encouraged to use civil liability law in order to be reimbursed by the liable person for the insurance sums they had paid to the insured. But as said, save for cases of gross negligence or intentional wrongdoing both the employer and co-workers of the victim-employee were freed from their civil liability against social insurance carriers or (subrogated) private insurance companies. Actions for reimbursement directed against other parties (manufacturer of goods, traffic members) were not subject to this condition but seem rather atypical in the field of occupational injuries.

5.4 The Belgian Widening of the Voluntary Industrial Accidents Insurance (1920–1970): Towards a 'Risque Social Insurance'?

As said in Belgium for victims of workplace injuries the early twentieth centu- 61
ry was marked by the introduction of a special act, the *Arbeidsongevallenwet*
1903 (Industrial Injuries Act, AoW) which made employers personally liable
towards the victim-employee, regardless of fault. This regime was not based
on any obligation to have insurance but rather rested on the employer's personal no fault liability (with the possibility of voluntary insurance). Similar to
the Dutch insurance coverage, this Belgian liability regime gradually expanded after the post-war years, for instance to accidents occurring on the way to
and from work. The risk of occupational disease was never brought under this
regime, but was shortly after covered by the so-called *Beroepsongevallenwet*
1927 (Occupational Disease Act, BoW). This 1927 Act offered compensation
for industrial employees and certain other professionals, viz. those with a recognized and known risk of occupational disease.

The voluntary accident insurance for employers was wide spread but started to raise serious complaints, which were mainly directed at its remarkably
low level of compensation: As said it covered no more than fifty percent of lost
income,[119] while the *Arbeidsongevallenwet* 1903 (Industrial Injuries Act,
AoW) barred the victim-employee's right to civil law proceedings. In 1930 the
amount of compensation was raised to a maximum of two thirds (66 percent)
of the victim's last wages, but this only applied to disability that lasted longer
than 28 days. For this first month of disability the damages remained at 50
percent. This was strongly criticized, especially given the fact that these
awards were also low compared to unemployment pensions and sick pay. As a
result the government further raised the disability pensions for the 'risque professionnel' in 1951, this time considerably to a maximum of eighty percent for
the first month which would after this month be followed by 90 percent of the
last salary. Employees who turned out to have been permanently disabled
through their work would then receive full compensation for their income
damage, albeit limited to a statutory ceiling. For the risk of additional income
loss, private insurance could be taken out. For some employees this was a collective private insurance taken out by the employer, which would then cover
the difference between the loss of the victim's actual income and the liability
awards.

Nevertheless this aspect was heavily criticized. Similar to English law, the
former view that the industrial injuries risk was a risk run and therefore shared
by both the employer (capital) and the employee (labour force), was losing
much of its appeal. The issue sharply resembled the Dutch position at the
time, albeit approached from the opposite angle: why, the Belgians asked,
would employees who ran such risks, be protected any less than they would
against the risk of other causes that threatened their level of existence? And
why would the obligation to pay for this have to be imposed on the individual

---

[119] See above, no. 41. M. Fontaine, La réforme sur les accidents de travail en Belgique et la distinction entre assurances de choses et assurances de personnes, in: J. Bedour (ed.), *Études
offertes à André Besson* (1976), 146.

employer? It would be better if these risks be borne collectively, by the (working) community as a whole. Based on this view and on the principle of solidarity, the system of social security gradually matured: Throughout the 1960-s most individuals were protected against the basic risk of medical expense and income loss regardless of its cause, but Belgium did *not* go as far as to abandon the distinction between 'risque professional' and 'risque social'.

62   Still, the government could not escape a gradual widening of the protection of employees suffering from workplace accidents. First it did so by extending the concept of industrial accidents; in 1941 the voluntary industrial accidents insurance had already come to include the risk of injuries driving to and from work. In the decades thereafter the circle of beneficiaries gradually widened and payments were raised. Also, comparable to Dutch law, special legislation was introduced to improve the position of state officials suffering from work related injuries (and occupational diseases).[120] But dissatisfaction continued to exist, not just with regard to the level of protection and the unequal position of workplace victims but also, or mainly, in respect of the fact that the no-fault regime made these victims completely dependent on the sense of responsibility and solvency of their employer. As said before, the *Arbeidsongevallenwet* 1903 (Industrial Injuries Act, AoW) did no more than that it imposed a personal obligation on employers against their employee to pay for the damages. This forced employees to, if necessary, start legal procedures against the employer. In later times the no-fault protection would become more 'real' than ever as industrial insurance was made compulsory by the *Arbeidsongevallenwet* 1971 (Industrial Accidents Act, AoW). This will be discussed in paragraph 6 below.

5.5   Minor Improvements Needed in German Law (1920–1970)

63   Even earlier than Dutch and Belgian law the scope of protection of workplace victims had been expanded in German law after World War I. The need for protection was felt strong due to the Post-war downfall and Germany too became subject to new social security legislation. Article 20 of the Grundgesetz (Basic Law) which was introduced in 1949 speaks of a democratic and socially responsible state. As a result of this employees in Germany are – by act of law – obliged to be insured and entitled to receive protection in cases of sickness or accidents, unemployment, in old age, or if they need care or suffer from a disability. They have to partly contribute to social security. These contributions, which are paid half by the employer and half by the employee, are based on four schemes of social insurance. For personal injuries the main two are the *gesetzliche Krankenversicherung* (Health insurance) and the *Rentenversicherung* (Pension insurance; for manual workers under the *Reichsversicherungsverordnung* (RVO) and for white-collar workers under the *Angestelltenversicherungsgesetz*). Then there are the *Pflegeversicherung* (Nursing

---

[120] *Act of 3 July 1967*, concerning compensation for occupational injuries, traffic accidents and disease of state officials. As said earlier, occupational disease compensation will be left aside here.

care insurance) and, of less relevance, the *Arbeitslosenversicherung* (Unemployment insurance). In addition to the existing social security schemes for medical health, disability and old age, social welfare and unemployment schemes were introduced in the mid-twenties. Based on the *Lohnfortzahlungsgesetz 1969* employers were made responsible for full salary payments during the first six weeks of sickness, as a result of which employees would keep their right to salary.

The growing demand for more state protection has had a strong impact on the basic mandatory *Unfallversicherung* (Industrial Accidents Insurance, UV) too, which by that time was ruled by Book 3 of the *Reichversicherungsordnung* (RVO), as it gradually became further detached from industrial or labour activities. First accidents which occur while travelling to and from work were included in 1925 (compare the current definition of *Arbeitsunfall* as given by § 8 subs. 2 of the *Sozialgesetzbuch*) as well as certain occupational diseases. More fundamentally, in 1942, the right to insurance was attached to individual employment instead of the (type of) business as such; it became a personal right of the employee. As the ideal of a welfare state furthered after World-War II, the basic mandatory *Unfallversicherung* (Industrial Accidents Insurance, UV) became concerned with the protection of civil defence workers, farmers, pupils, students and pre-schoolers in kindergarten. Even emergency rescuers, people helping at the scene of an accident, blood and organ donors and in some voluntary activities the volunteers were included (the so-called *unechte Unfallversicherung*). Separate industrial accident funds for the public sector were also put in force for the benefit of *public officials* (employees or wage earning workers in the public sector). Next to this public system of statutory accident insurance, German law has a private-law accident insurance scheme. For this growing number of insured risks therefore the employer was free of civil liability actions except for where the public law benefits were made due to *intent or gross negligence*. 64

Both the basic mandatory *Unfallversicherung* (Industrial Accidents Insurance, UV) and social security legislation shared the same goal: protection of the working class by securing living standards based on income related benefits in cases of sickness, unemployment and retirement. The industrial accidents insurance was and still is for the best part, financed by the employer exclusively however, while social insurance was paid for by social insurance contributions[121], similar to Dutch law (albeit the Dutch legislator had traded the industrial accidents insurance for social insurance). The contribution charged for the accident insurance was, and still is, not standardized: It was made dependent not only on the total sum of annual pay of benefits but also on the accident branch ('*Gefahrenklasse*') of the given business. This means that the contribution would, and will, be higher if the risk of accident for the employees in 65

---

[121] *Not* through general taxation, see M. Seeleib-Kaiser, The Welfare State: Incremental Transformation, in: S. Padgett/W.E. Paterson/G. Smith (eds.), *Developments in German Politics 3* (2003), 142.

the employer's enterprise is (statistically) higher. Not only does this promote
that the contributions will reflect its good health and safety practices but also
it gives incentives to take the precautionary measures that are needed to reduce the risk of accidents. Various other regulations, safety measures but also
recovery rules and occupational integration assistance and alike were adopted
in order to promote damage reduction, for which the executive accident insurance bodies, the *Berufsgenossenschaften* (BGs), see no. 13 above were made
responsible. Under the *Unfallversicherungs-Neuregelungsgesetz* (UVNG), the
new Industrial Insurance Act which was introduced in 1963, the BGs were
granted the power to sanction violations of rules on prevention and alike. As a
result even today the German insurance system shines in prevention and rehabilitation, see further the contribution by Philipsen. These executive bodies
were organised in three groups based on branches of production. The first
group is formed by the industrial employers' insurance associations, each of
which covers a sector of industry. A second group embraces the agricultural
Berufsgenossenschaften (BGs). Both groups are, aside from a few exceptions,
organised on a regional basis and funds in the agricultural sector also receive a
federal subsidy. The third group of *Berufsgenossenschaften* (BGs) deals with
the public sector. Here the accident insurance bodies are responsible for public
sector employees, children, pupils and students attending educational estab-
lishments, voluntary workers and others.

5.6   Comparative Notes

66   The turn of the nineteenth century was followed in all four law systems by the
introduction of social security legislation, either in addition to or as a replacement of the system of industrial accidents insurance. In the years that followed
there was a clear trend to extend the special protection against the 'risque professionnel' to *all* risks of injury or death, regardless of their cause. The – coin-
cidental – manner in which the damage was caused was not found significant
for the level of compensation: why should one who falls at work be treated
different from one who slips in his own bathroom? It is the nature of the risk
(threatening the individual's existence) that should be decisive for the scope of
protection, not the way this risk had materialised.[122] These risks, if run by employers, either caused by workplace injuries or otherwise, should be carried by
the business as a whole, every individual whose protection was included
would have to contribute, and not just their employer.

Around the fifties social security legislation was introduced for basic risks
of personal injury. In the Netherlands this development resulted in a retreat
from the distinction between workplace and other victims in 1967. This was
contrary to many other West European countries such as Germany, the UK
and Belgium. These latter three law systems shared in common that they offered a special statutory insurance for industrial accidents. In German and English law this was an integral part of social security. In Germany it was financed by contributions of the employer, in English law by contributions of

---

[122]   See e.g. for Dutch law *Kamerstukken II,* 1962–63, 7171, no. 3, 2.

both employers and employees (here, premiums were made independent from the nature or gravity of the risk they covered) and by subsidies by the State (through taxes). The scope of protection of the German Unfallversicherung (Industrial Accidents Insurance, UV) was much wider than the English national industrial insurance's, not only with regard to the categories of insured persons but also given the kinds of damages that were covered (in English law only the loss of income). Better yet, in Belgium and England the level of protection seemed relatively modest: only the loss of income (England), a mere half of the victim's damages (Belgium).

Dutch law now differed, and still does, completely from the remaining law 67 systems as it rejected the distinction between damages caused by work related injuries and other causes altogether. The growing believe was that certain risks (such as high medical costs, disability, unemployment, old age etc.) are run by all individuals regardless of their cause. Such risks deserved public law protection which would be best carried by the public at large. Medical expenses, income loss and damages through death were now covered by general social security – regardless of their cause. Contributions for basic protection were taken from all inhabitants, as well as from employers, employees *and* the State, based on the concept of solidarity. In England too the social security system was built on this principal idea (combined with additional protection from civil liability law), but there victims with work related damages were, as said, given a more preferential treatment.

In both Dutch and English law the gaps that social security awards left were filled by liability law.[123] Here, victims of workplace accidents were offered extra protection: The liability regime for workplace accidents was in each of both law systems more protective than the general rules of liability law were. As a result in both systems the financial risks of workplace injuries were wider spread: Tax payer, employees and employers were paying, whereas in German and Belgian law the financial burden was carried exclusively by employers.

6. Shift III (1970–2004): Social Reform and Private Compensation Schemes

6.1 The Need for Social Reform

After the introduction of social security, working conditions and safety regula- 68 tions with regard to the workplace would only be improved.[124] Next to the introduction of the European Social Fund in 1960, which needed to promote employment conditions in the continent, anti-poverty actions and regulations on the equal treatment of men and women throughout the mid-seventies, 'safety and health in the workplace' became a central theme of European so-

---

[123] P. Cane (*supra* fn. 42), 275.
[124] In the Netherlands for instance, the Minister of Social Affairs Veldkamp who was responsible for the abolition of the OW estimated that the 190,000 persons relying on payments based on the OW would remain stable after the introduction of the Occupational Disability Act (WAO) in 1967 (150,000 to 200,000).

cial policy.[125] The ideal of a welfare state came to include the introduction of a system securing 'aid and other means to prevent, repair or compensate personal damages'.[126] Offering medical aid and compensation of health damages were high on the political agenda[127]. The working conditions in factories, training of employees and alike had indeed improved over the years and as a result the risk of injuries was expected to stay within reasonable limits.

69   But the need for social security payments did not slow down. On the contrary, the social security system expanded largely, amongst other factors due to other causes of disability and illness as well as – to a great extent – certain forms of abuse of the system.[128] Economic factors such as the dramatic rise of oil prices and the increase of labour costs, compared to the new and expanding Asian market and the rising cost of health care in the mid seventies, added to high numbers of unemployment. The 1960-s and 1970-s were marked by a great rise in public spending on social security benefits and an expansion of the welfare state in all four law systems. As a result the idea of a welfare state gradually started to deteriorate. John Maynard Keynes' paradigm of spending policy in times of unemployment and inflation now seemed to fall short. The growth of public expenses concerning the economical, social and financial position of its citizens was challenged. In continental Europe rising social insurance contributions and tax rates were expected to undermine the international competitiveness of national companies. Changes regarding the manner in which the welfare state was financed were found necessary to gain economic improvement.[129]

70   This also convinced the authorities that injury reduction and loss control could have a significant effect on costs. It strengthened the idea that full compensation had a price and should not discourage the worker's personal responsibility.[130] Despite these political views a few basic rights that the European Community Charter granted to citizens promoted a safe and healthy workplace, such as the right to improvement of living conditions and working conditions

---

[125] H.G. de Gier/P.J. van Wijngaarden/A.M.E. Roelofs (*supra* fn. 15), 24 *et seq.*

[126] Translation from J.G.F.M. van Kessel, *Sociale zekerheid en rechtsbeleid* (1985), 1 where social security is defined as: 'de toestand van de maatschappij, waarin ieder lid van de samenleving verzekerd is van hulp en middelen gericht op voorkoming, herstel of vergoeding van menselijke schade'.

[127] H.G. de Gier/P.J. van Wijngaarden/A.M.E. Roelofs (*supra* fn. 15), 32 and 40.

[128] For Dutch law R. van Veen/D. Bannink/R. Pierik/W. Tromme, *De toekomst van de sociale zekerheid* (1996), 62 argue that '[t]he problem with the Occupational Disability Act (WAO) is not that in the Netherlands so many persons are unable to work. The problem is that too many are diagnosed with occupational disability. As early as the 1970-s it became clear that based on doubtful grounds or arguments – and even in violation with statutory rules – disability payments have been made' (my translation, EE).

[129] See more extensively on the political determinants for social change a. others M. Seeleib-Kaiser (*supra* fn. 121), 143 *et seq.*

[130] Also in legal doctrine there is support for this view, see *e.g.* J.D. Carr (*supra* fn. 11), 420 who claims statistical evidence indicates that employers or co-workers are legally responsible in whole or in part for only about 25 percent of all industrial accidents and that less than a full wage replacement concept will encourage workers to return to work as soon as possible.

and the right to social protection.[131] Yet the ultimate goal of some of the subsequent compensation programs at national level, many of which are still operative and improved, was to make safety a value (see in this respect the contributions of Philipsen and Hoop).

Particularly the public expenditure on income replacement in the case of long-term or permanent disability of employees increased. To turn this process, national government's social policy came to involve employers more actively. As a result employers were made subject to many health and safety regulations and made financially responsible for the risk of disability pensions. Specific regulations on keeping a safe and adjusted work environment for each of their employees were put in force, providing access to treatment at an early stage, and administering the disability rate in the individual employer's business. At the same time there was concern that incentives to reduce injuries would cause employers to underreport injuries.[132]

6.2   The Current English Compensation Scheme (1970–2004)

6.2.1   Initial Basic 'Risque Social' Protection

This new social policy also affected the position of employees with personal   71
damages caused by a workplace injury or occupational disease as the costs of workplace accidents were growing. In the late 1960-s the Pearson Committee published statistics showing that every year some 1,300 people were killed and over 700,000 were injured at work. In those years the social security system became subject to strong criticism, the main point being that the government's flat-rate system was incapable of providing a minimum standard of living for its citizens. Consequentially in 1966 sickness and industrial injury benefits as well as unemployment benefits were made earnings-related, similar to retirement pensions. In 1975 the contributions to those systems were related to the contributor's income too. Preventive measures were taken to reduce these costs similar to the current Health at Work program for employees in the National Health Service and occupational health and safety measures (*e.g.* as laid down in the Factories Act).

These damages, if caused by industrial accidents, were, and still are, primarily   72
a concern of social security law. As was seen in the previous paragraph the National (Industrial Injuries) Insurance Act 1946 introduced a special social security scheme for earnings related benefits caused by workplace accidents or occupational diseases. These benefits were paid by the state and financed through contributions to a common fund, collected from employers and employees and a subsidy by the Exchequer. The risk of medical expenses (and other personal risks such as old age, maternity and alike) was financed exactly the same way but for this equal protection was offered for all accident victims, governed by the general scheme of the National Insurance Act 1946. While, in

---

[131]   H.G. de Gier/P.J. van Wijngaarden/A.M.E. Roelofs (*supra* fn. 15), 29.
[132]   T.L. Guidotti/J.W.F. Cowell (*supra* fn. 11), 447.

other words, for medical expenses no distinction was made, for earnings related damages employees who suffered from workplace accidents or occupational diseases were treated more favourably than other victims. This more favourable scheme based on the National Insurance (Industrial Injuries) Act 1946 was made possible as industrial accidents on the whole were still a minority cause of disability, which meant that for relatively small contributions relatively high benefits could be offered. The Act took compensation out of the hands of employers and tribunals and gave employees clearly defined universal rights.

In 1980 however, the so-called White Paper proposed to shift the responsibility for the administration of short-term sickness payments from the state to employers as they were expected to deal with this more effectively and already had sickness schemes.[133] Based on this the Social Security and Housing Benefits Act 1982 first introduced the employer's obligation to Statutory Sick Pay (SSP). Originally he was made responsible for short term disability: for the first eight weeks of disability due to sickness or invalidity and in 1992 extended this to a good six months (see nos. 73 and 74 below). The employer could recover in the form of rebates on the national insurance contributions but in 1994 the employers' national insurance rebate was abolished, except for smaller firms.[134] The first three 'waiting days' were left unpaid.

73   By more recent legislation, based on part XII of the Social Security Contributions and Benefits Act 1992, the employer was made responsible for Statutory Sick Pay for the first twenty-eight weeks of mental or physical disability, with the exception of the first three days (s. 155 Social Security Contributions and Benefits Act 1992). This applies to all employees who were under a contract of service and were incapable to work 'by reason of some specific disease or bodily or mental disablement'.[135] After those first twenty-eight weeks all are entitled to subsequent social security benefits. The Social Security Contributions and Benefits Act 1992 offers unemployment benefits, retirement pensions, child's special allowance and other allowances, but also benefits in cases of sickness, invalidity and death (financial dependents). Special benefits are offered to employees who suffer from workplace injuries. Below, this special scheme for income loss caused by the 'risque professionnel' will be discussed in more detail. With regard to the risk of medical expenses the victim-employee automatically lapses into the general scheme of social security. Here, workplace victims are in no better position to recover than employees who suffer from a natural cause or non-workers. The National Health Service has been given the task to provide medical and social care, financed by the State (by raising taxes). In other words, as far as the costs of health care are concerned, regardless of whether they were caused by work related injuries, the taxpayer effectively pays for them.

---

[133] *Income During Sickness: A New Strategy*, Cmnd. 7684, 1980; referred to by S. Deakin/G.S. Morris (*infra* fn. 138), 356.
[134] S. Deakin/G.S. Morris (*infra* fn. 138), 356.
[135] Social Security Contributions and Benefits Act 1992, s. 151 (subs. 1 and 4) and 155. S. Deakin/G.S. Morris (*infra* fn. 138), 357 mention that the weekly SSP rate was set at 55.70 Pounds from April 1997.

6.2.2 Special Social Insurance for Income Loss Due to 'Risque Professionnel': Industrial Injuries Scheme (IIS)

As said the Social Security (Contributions and Benefits) Act 1992 (and the 74
Social Security Administration Act 1992) had serious ramifications with respect to the recovery of *income loss* suffered from a workplace accident or occupational disease.[136] Part V of the Act ('Benefit for Industrial Injuries') contains special rights and regulations applicable in the case of industrial
accidents or one of over seventy industrial diseases known to be a risk from
certain jobs. This special section, the so-called Industrial Injuries Scheme,
hereafter: IIS, was introduced to govern long-term incapacity to work which
was caused this way and is part of National Insurance. The IIS is regulated by
public law and carried out by the Department of Social Security. It is financed
through contributions of employers and employees, apportioned to income
(and not to the accident rate of the branch of industry).[137] It only protects employed earners who are under a contract of service or are employed as office
holders such as company directors (hereafter: the injured employee) and different from Germany, law trainees and alike are excluded. In a – limited – way
employers are thus made responsible for the risk of income loss due to work.

In short, the IIS puts the injured employee in a slightly more favourable
position, compared to those whose damages were not caused by work related
injuries. As was said before, the employer is held responsible for Statutory
Sick Pay for the first twenty-eight weeks of disability, regardless of its cause.
Subsequently, the industrial accidents scheme offers relatively high disablement benefits for the injured employee if he is physically or mentally disabled
by at least fourteen percent, and further it offers reduced earnings allowances,
retirement allowances and industrial death benefits as well as an increase of
the disablement pension if the victim is fully disabled and requires constant attendance (s. 94 (2) and 104 (1) Social Security Contributions and Benefits Act
1992). Many employers make payments of industrial sick pay on top of these
social security benefits.[138] The main condition is that the employee suffers
from a personal injury which was caused 'by an accident arising out of and in
the course of employment'. If the accident occurred in the course of employment it is presumed to have arisen out of that employment in the absence of
contra evidence, says subsection 3 of section 94 of the Social Security (Contributions and Benefits) Act 1992 (s. 29 and 30 of the Social Security Act 1998
give extra rules on the decision whether the accident was or was not an industrial accident). This includes commuting accidents and accidents that are
caused by either another's misconduct or negligence or by being hit by something, provided this happened in the course of employment (and where commuting accidents are concerned with the exception of 'normal' accidents with
public transport). In all cases the injured person must have been 'in the course
of employment' at the time of his injury and, cumulatively, the injury must be

---

[136] Which includes state officials, see B. Barentsen (*supra* fn. 87), 129.
[137] The contributions were and are made to a common fund as collected by the Contributions
Agency – HQ and local offices.
[138] See S. Deakin, S/G.S. Morris, *Labour law* (1998), 354.

the result of risks peculiar to his employment. A car crash or a robbery can well be related to the injured person's employment in the sense that it happened during working hours but may not be peculiar to his job.

75 Interestingly though, the injured employee will lose his claim to these special awards if at the time of the accident he was acting in contravention of any regulation applicable to his employment or any orders given by his employer, unless it is showed that the accident would have arisen anyway *and* that the employee breached the order or regulation for the purposes of (and in connection with) his employer's trade or business (s. 98 Social Security Contributions and Benefits Act 1992). Opposite from the rules of tort liability, contributory negligence will thus make the injured employee lose his entire claim based on this more preferable scheme.

What is more, the IIS only covers a relatively small part of these damages: Its payments were not meant to compensate income loss but to compensate the employee's incapacity to work, with the result that only his degree of invalidity plus certain other personal circumstances such as his age and sex are taken into account. As a result payments were not made dependant on the employee's true salary. Furthermore, only incapacity to work lasting longer than fifteen weeks will entitle the employee to payments while, reputedly, more than 60 percent of all victims of workplace injuries return to work within three days.[139]

76 Considering all this, the IIS has been subject to great criticism. First because of the limited amount of extra protection that it seems to offer compared to the protection for long-term incapacity to work as offered by the 'regular' (*i.e.* 'risque social') social security scheme. Secondly, it covers the risk of injury 'arising out of and in the course of employment', which is said to be difficult in at least two ways.[140] Some deny the soundness or justification ground of this criterion as a distinguishing factor. In their view there is 'no clear and sound policy reason for distinguishing between employment risks and non employment risks; and so it is almost impossible to construct a satisfactory criterion for distinguishing injury within the scheme from injury outside the scheme; this generates many borderline cases'.[141] The criterion that the injury must have arisen 'out of employment' too has been the object of criticism: Why, some authors ask [142], should industrial accidents have to be treated any different from accidents that occur outside working hours?[143]

---

[139] B. Barentsen (*supra* fn. 87), 130.

[140] P. Cane (*supra* fn. 42), 282.

[141] P. Cane (*supra* fn. 42), 282.

[142] P. Cane (*supra* fn. 42), 283: 'Certainly we appear to have got very close to a point where an accident arising in the course of employment will almost inevitably fall within the system, and be treated as having arisen 'out of' the employment. This makes it difficult to justify having a special scheme for work-caused accidents. But these provisions have, at least, had a good practical effect: In 1976 a Social Security Commissioner commented that the 'out of' requirement now gives little trouble'.

[143] Despite this criticism it has remained to be in force.

6.2.3 Additional Protection from Civil Liability Law

In concurrence with liability law however, it is the liable person who effec-   77
tively needs to pay for most damages, either insured or uninsured. As for social insurance, statutory sick pay and the IIS as discussed above, the liable
person (the employer, a third party or their liability insurance company) will
be sent the 'bill', based on the Social Security Administration Act 1992. This
is not limited to work accidents and now ruled by the Social Security (Recovery of Benefits) Act 1997. Based on this ruling liable persons must re-pay in
principle *all* social security benefits that were awarded to the victim. Recovery by the state is limited though to benefits paid in the first five years after
the accident or the date when the claim is determined (settled). For that purpose, section 82 rules that the liable person (or his liability insurance carrier)
shall not compensate the victim before the Secretary of State has furnished
him with a certificate of total benefit (and if the liable person *himself* requests
the Secretary of State to furnish him with a certificate, the Secretary must in
principle send him one within four weeks after having received his request, s.
94). Based on this certificate the liable person or his insurer must deduct the
corresponding social security benefits from the amount of compensation
owed to the victim (*e.g.* earnings related benefits must be deducted from the
part of the liability sum that is meant to compensate the victim's income
loss). For these amounts the victim-employee loses his entitlement to compensation from the liable person, as he is entitled to the IIS benefits (s. 82 (2)
rules that his right to compensation shall be regarded as satisfied to the extent
of the amount certified in the certificate of deduction; similar to the position
of Dutch victim-employees, see no. 85). As far as the victim has right to
compensation for any other – uninsured – kinds of damage, such as pain and
suffering, this right to compensation can*not* be reduced in any circumstances.
Damages for non-pecuniary loss are 'ring fenced' – i.e. the claimant gets
those in full from the defendant, whereas in the case of income losses and expenses the defendant deducts the amount of the benefit from the damages and
pays that to the state. But with regard to non-pecuniary losses this benefits
the claimant, not the defendant, who still has to pay the state an amount
equivalent to all benefits received. There is no offset for the defendant for
contributory negligence. The defendant needs to pay the state the full benefits. As said the state's right to recovery is limited to benefits paid in the first
five years.

   Next to this reimbursement scheme, which makes employers or third parties, as liable persons, pay for the part of the victim's damages that fall under
social insurance, the former also remain liable towards the victim-employee,
for damages that are *un*insured. My guess would be that the restrictive governance of the IIS could force employees into civil litigation. More surely, as a
result of the limited amount of protection offered by the IIS income losses are,
for the most part[144], left uncompensated, which means the employee is made
dependent on either private first-party insurance or on a civil liability action.

---

[144] See above and B. Barentsen (*supra* fn. 87), 131.

As for the latter, the employer's civil liability is still based on personal wrong-doing and fault, which must be shown by the claimant (victim-employee). Many occupational health and safety statutes and rulings however impose regulatory duties upon employers which are enforced through the powers of the health and safety inspectorate and ultimately the sanctions of criminal courts. In some cases of a breach of a criminal statute the victim-employee may bring an action for damages via the tort of breach of statutory duty. In the absence of a specific statute the employer may be held liable in negligence for a breach of the general common law duty to provide safe working conditions.[145] This common law duty is taken to include the duty not to expose his employees to a stressful working situation or workload.[146]

Furthermore, European directives and more specifically those implemented under the so-called Health and Safety at Work Act 1974 as well as section 44 (and for Sunday workers section 45) of the Employment Rights Act 1996 impose a duty for the employer to ensure 'so far as is reasonably practical' the health, safety and welfare of his employees.[147] Under the 1996 Act the employer is exonerated insofar as he shows that the injured employee acted negligently in such a way 'that a reasonable employer might have treated him as the employer did' (subs. 3 of s. 44). However, these regulations do not give rise to liability but are merely of a regulatory nature. Reputedly nowadays the English face, at the outside calculation,[148] annually 115,000 tort claims for industrial injuries of which about 78 percent seemingly result in some payment of compensation. At the very least 3,000 of those claims are said to be for occupational diseases (RSI, stress and alike) of which 57 percent seem to result in compensation payments.

6.3   The Current Dutch Compensation Scheme (1970–2004)

6.3.1 The Need for Social Security Reform

78   In the Netherlands to this day most personal damages caused by workplace injuries or in any other way will be governed by social security programs and private insurance. The risk of basic medical expenses will generally be covered by private medical insurers while 'exceptional' expenses such as long-term hospitalization are covered by social insurance. The risk of income loss will for the first two years (short term disability) be taken care of by the employer and/or by his private insurance company. After two years the victim will lapse into social security (and for additional income damage he may have taken out private insurance or will have a civil liability action).

---

[145] See more in depth S. Deakin/G.S. Morris (*supra* fn. 138), 319 *et seq.* and B.S. Markesinis/S. Deakin/A. Johnston (*supra* fn. 41), 560 *et seq.*
[146] See S. Deakin/G.S. Morris (*supra* fn. 138), 323, referring to the Court of Appeal's decision *Walker v Northumberland County Council* [1995] *IRLR* p. 35.
[147] S. Deakin/G.S. Morris (*supra* fn. 138), 319 and 323 *et seq.*
[148] P. Cane (*supra* fn. 42), 179 makes mention of the figures given but argues 115,000 is 'probably now rather too high given the fall in the last thirty years in the numbers of people killed and injured at work'.

This current system has had a long history of reform. Similar to the English approach, the Dutch government has been forced to take action as the need for drastic social security cuts grew stronger from the early eighties onwards. In those years the Christian-'right wing' government (CDA-VVD coalition) started its new policy based on the view that the welfare state had not been efficient and that it could no longer afford its costs. Not only was the ageing population an important factor which contributed to this but also it seems that in the Netherlands there was a relatively high demand for social security – mainly due to higher accessibility.[149] An essential part of the government's approach to turn this process around was to make employers face responsibility for this growing public expense by making them bear the financial risk for the short-term and long-term disability rate in his (individual) business. In 1992 this concern of reducing the disability rate led to the introduction of the so-called 'bonus-malus' system. This system made the employer risk a sanction in the form of a fine (*malus*), for any new employee who was reported disabled, unless the employee would be reintegrated elsewhere in the employer's business.[150] For (partly) taking in disabled employees, employers would make a gain in the form of a subsidy (*bonus*). The system was under serious attack however. This came for an important part from the fact that the 'malus' that was risked by employers was made irrespective of the exact cause of disability. Why did the employer have to pay for personal risks run by his employees in their private lives such as going on a skiing trip, the employers' lobby argued (and this objection was also heard in Parliament[151]), while such events are typically beyond his control? Partly as a result of this objection (directed against its 'risque social' nature) the bonus malus system was abolished in 1995, but within a few years' time a very similar system would be introduced.

This new system, based on the so-called *Wet premiedifferentiatie en markt-* **79** *werking bij arbeidsongeschiktheidsverzekeringen (the Disability Insurance (Differentiation in Contributions and Market Forces) Act, Wet Pemba)*, made the social contributions that employers owed to finance the (former) *WAO* dependent on the disability rate of each employer in his individual business. Was his disability rate above the average, this would raise the contributions which were collected from him. Different from the bonus-malus system was that this was decided by looking at the average disability rate, which made it less likely that the individual employer's contributions would go up if the disability rate in his individual business had gone up as a result of private circumstances not related to work.[152] Making the average disability rate decisive made the 'ris-

---

[149] In this sense M.W. Dijkshoorn, Ontwikkelingen in de WAO, [1996] *Verzekerings Archief (VA)*, 82–86 and likewise G.K. Einerhand/R. Prins/T.J. Veerman, Ziekteverzuim en WAO in enkele Westeuropese landen, [1995] *Sociaal Maandblad Arbeid (SMA)*, 520–528.
[150] *Wet terugdringing arbeidsongeschiktheidsvolume* (Act Reducing Disability Rate), Kamerstukken II, 1990–91, 22 228 and *Wet terugdringing beroep op de arbeidsongeschiktheidsregelingen* (Act Reducing the Demand for Disability Benefits), Kamerstukken II, 1992–93, 22 824.
[151] Kamerstukken II, 1991–92, 22 228, 612–613.
[152] P.S. Fluit/A.C.J.M. Wilthagen (*supra* fn. 82), 120.

que social' nature of the financial burden for employers more acceptable. As Philipsen shows in his contribution, in the years that followed the disability numbers have decreased significantly (which might mean that there is a correlation with this internal shift in governance).

80   Additionally in 1996 a ruling was put in force which required employers to continue to pay at least seventy percent of their employee's last income throughout the first year of his work related disability (the exact percentage is made dependent of the individual worker's work related disability). This labour law duty may not be circumvented by letting the employee off; prohibitive clauses regarding redundancy seek to protect the latter during the first two years of his incapacity to work. In 2004 the Dutch government extended this duty to continue salary payments to two years (art. 7:629 Civil Code). The risk of disability due to pregnancy and a minor category of other sources of disability is borne by the employer and employees collectively, based on the *Ziektewet 1913* (Sickness Benefits Act, ZW).

By imposing this obligation to continue salary payments in the absence of the work being done, employers are given a strong incentive to prevent accidents or the deterioration of the harm done. It must be noted though that this labour law obligation to continue salary payments is independent of the cause of the disability. The idea is that, even in the – high – number of cases where the disability is *not* work related, the employer is still the better party to promote fast recovery (*e.g.* by adjusting the working environment or the time schedules in such a way as to make sure that the disabled employee is put in a position which will enable him to resume working on full capacity).

81   The victim-employee who is still found incapable of resuming work after this first period, in which the employer has continued to pay his wages, lapses into social security. Again this is independent of the way in which the damage came about. The (former) WAO offered income related pensions for employees who suffered from sickness or disability that had lasted more than one year, regardless of its cause; the exact amount of these pensions was and still is dependent on the disability rate that was applicable in his situation. This system was paid for through the contributions of both the employer and his employees. But more recently, under the (former) central-right wing government, the legislator has introduced a new arrangement for the risk of income loss after the first two years of income related payments that has replaced the WAO on 1 January 2006. Based on this new social security legislation the right to income replacement payments after the first two years of disability (in which the employer must continue to pay wages, see no. 80) has been limited to those who have no prospect of resuming work. Now, employees will only remain to be entitled to these payments if they are fully and permanently disabled to work. Those who are partially disabled are given a strong financial incentive to resume working again: If they do not resume work they risk to be left with nothing more than unemployment benefits, related to the minimum level of income (as opposed to the former regime which compensates their income loss to 70 percent). Only once they do resume working will they be giv-

en completion of these payments. The idea behind this new legislation seems clear:[153]

> 'By focusing more on preventing and controlling the absence from work a better work environment will be created, which will promote productivity. The numbers of absence from work and employees depending on WAO are declining but nonetheless serious social reform in this respect seems necessary. The proportional increase of the ageing population involves an immense pressure on the Occupational Disability Act (WAO) (...)'.

In short this new system, based on the so-called *Wet Werk en Inkomen naar Arbeidsvermogen* (Work and Income Ability Act, WIA), only offers income replacement pensions to those who are disabled by at least 35 percent. Like the (former) WAO the system does not care how the work related disability came about (work related or private accident); it covers the 'risque social'. It introduces two different schemes: 82

1. the *Regeling Inkomensvoorziening Volledig Arbeidsongeschikten* (Pension Scheme for Fully Disabled Act, IVA) offers income related payments, for those employees who are disabled permanently and for 100 percent, and
2. the *Regeling Werkhervatting Gedeeltelijk Arbeidsongeschikten* (Back to Work Scheme, WGA), which sees to the continuation to work for those partially disabled to work.

The first scheme makes sure only those who are fully (that means 100 percent) and permanently incapable of working will receive and continue to receive, income replacement benefits. The latter scheme offers 70 percent of the lost income to those who – out of 39 weeks prior to their disability – had been working for at least 26 weeks. These payments will, dependent on the employee's period of disability, be continued for at least 50 percent of his remaining capacity to work; others will be left with 70 percent of the minimum wages times the percentage of their disability.

A brief note must be made with regard to the self-employed, that is: all entre-   83
preneurs, tradesmen, businessmen and alike, even though this category of professional workers falls outside the scope of this book. The reason for this is that an important shift of governance took place for persons in this category who due to their handicap are not able to keep their prior earning capacity. While in recent years, before August 2004 special social security legislation used to offer the self-employed an earnings related pension for situations in which their incapacity to work had lasted more than a year, now the self-employed are expected to have obtained private insurance on a voluntary basis. Except for welfare programs, there is currently in principle no social security scheme whatsoever, although for those who cannot obtain private insurance for health reasons or certain other reasons the state provides for an alternative form of insurance.

---

[153] Brief van de Minister van Sociale Zaken en Werkgelegenheid, 16 september 2003, *op. cit.* (transl., EE).

It seems likely that particularly this category of victims will seek (additional) protection from liability law in case they get injured. Although the self-employed typically have no employer, a small increase of employers' liability claims could perhaps be expected for those cases where the victim had both a service agreement and – in free hours – his own business and suffered from a workplace accident in the course of his (service) employment. He may then sue his employer based on civil liability law not only for his lost earning capacity in his quality as employee but also for his lost profits as entrepreneur. The latter part of his damages (lost business profits) will be more substantial now that social security protection is generally missing.

6.3.2 Gaps in Social Insurance Filled by Other Compensation Systems

84 In addition to the labour law arrangements and social security benefits as aforementioned, Dutch liability law allows the victim-employee to sue his employer (and/or the – voluntary – liability insurer of the company) for damages. Also he may claim damages from his own private medical insurance company. The continuing salary payments and special social security benefits to which he is entitled will in both instances be deducted. Based on special rules of liability law the amount of benefits that the victim will receive from his employer and social insurance carriers will be deducted from his right to compensation for the same damages.

As for the latter, the risk of personal damages due to workplace accidents will in many cases be covered by individual private sickness or invalidity insurance schemes or in collective insurance arranged by the employer. Changes of social policy may then easily have side-effects for both these compensation systems. This seems especially likely since 1994, when a coalition of liberal and socialist parties (VVD-PvdA) put privatisation high on the government's political agenda. Along with this new policy came the rehabilitation of the idea of personal responsibility. Drastic cuts in social security pensions for widows and other financial dependants of the deceased, in the case of death, were argued by referring to the personal responsibility to maintain a living or at least take out private life insurance. Similar arguments were used for cuts with regard to income replacement payments for the long-term disabled, the WAO. All this may have created a higher demand for additional private insurance against the risk of (additional income damage), and/or a new market for private insurance.[154]

85 Assumingly[155], social cuts have also been (partly) responsible for a gradual expansion of the system of employers' civil liability, or at least for the growth of

---

[154] E. van de Beek, Kassa voor verzekeraars. Het kabinetsbeleid pakt voordelig uit voor de verzekeraars, die flinke omzetten kunnen maken. Grote verliezers zijn pensioenfondsen en WAO-uitvoerder UWV, [2004] *Elsevier,* 72–73 (25 September 2004) claims this has clearly worked for the benefit of these private insurers (*'de 'WAO-hiaatverzekering' werd een populaire arbeidsvoorwaarde'*).

[155] See B. Barentsen (*supra* fn. 87), 18. Surely these assumptions are causative in nature and therefore hard to quantify and to support by empirical numbers. Still it is fascinating to see that while cuts were made in the social security system the civil law rules as to employers' liability were interpreted more strict than ever.

civil law actions for damages *vis-à-vis* the employer. He is found best equipped to control his employees and give instructions regarding the way the work must be done. As was seen above there has always been a strong connection between Dutch social security and civil liability law. Originally, lacking any social security scheme, injured employees or their financial dependants relied on private initiatives of poor relief and in a few cases civil liability procedure. The first Dutch public law insurance with social security characteristics, the OW, limited employer's liability to – mainly – intentional wrongs and claims for property damage. For damages that were covered under the old regime the injured employee lost his right to a civil action against the employer and fellow employees. The insurer was subrogated in his claim. Ever since the OW was repealed, in 1967, employers risk liability against the employee, for damages that are not yet compensated by social security benefits.[156] But if the employer by his intent or gross negligence caused the damage he might also be held liable to repay the social insurance carrier, similar to English law (see no. 77). To the extent of social security benefits the victim-employee will lose his entitlement to civil law compensation *vis-à-vis* his employer (as in English law this is independent of whether the social security benefits have already been actually paid by the social insurance carrier and received by the victim). Again similar to English law the social insurance carrier may seek reimbursement independent from the victim. It must be noted though that this action for reimbursement is limited to wrongs by the employer or a colleague that were due to their respective intent or gross negligence. Aside from this aspect, the rules for establishing employers' liability based on an action for reimbursement by the victim's insurance company are the same as they would be in a direct action for damages (from the victim).

To conclude, the abolition of the OW thus made the employer the paying party on an individual basis, based on civil liability, statutory sick pay and his obligations to contribute to social schemes for long-term disablement awards. Of course this does not mean that it is the individual and not the public at large who pays for liability claims. Nor does it mean that insurance companies do not take part in the system. Barentsen actually argues a more recent 'shift' within Dutch liability law in the sense that employers' third-party insurance tends to cover injury risks that used to be covered by the social security scheme.[157] At any rate it seems likely that the development and growth of civil law rules imposing liability on employers has had an effect on the availability, price and rate of liability insurance. Liability insurance alternates social security or private insurance schemes by covering risks that would otherwise be left uninsured. Distinct from social and private insurance schemes the liability

86

---

[156] Employers' liability for damages that were caused at the workplace can be derived from both labour law (art. 7:658 of the Dutch Burgerlijk Wetboek (Civil Code, BW) imposes a strict liability on employers for hazardous workplace, machinery and alike, while article 7:611 BW contains a duty to act as a good and reasonable employer) and tort law (art. 6:162 *et seq.* BW). See extensively S.D. Lindenbergh, *Arbeidsongevallen en beroepsziekten* (2000), 11 *et seq.*

[157] B. Barentsen (*supra* fn. 87), 15.

system offers full compensation, which means that all damages that are left uninsured must be repaired by the liable person or his insurance company.

87   While strictly speaking still fault based, in case law the employer's liability has developed into a rather strict ground for compensation. Actually two special grounds for liability vis-à-vis his employees can be found in labour law, as laid down in Book 7 of the Dutch *Burgerlijk Wetboek* (Civil Code, BW). In 1997 the aforementioned article 1638x BW (see no. 31 above) was replaced by article 7:658 BW, which still aims at violations of the employer's duty not to expose his workers to an unsafe workplace or to hazardous working conditions. Article 7:611 BW imposes a more general duty to act as a good employer which can be invoked by workers to hold their employer liable, provided specific conditions have been met. The requirements for liability are that the employer acted negligently (*i.e.* he violated the aforementioned duty) and that the damage was caused in the course of employment. The employer's responsibility for the work environment and working conditions *vis-à-vis* his employees aims at the protection of his employees against the risks inherent to their employment. As said he is in the best position to give instructions and to decide how, in what speed and under which circumstances the work must be done.[158] Furthermore, bearing the risk of responsibility must encourage employers to take precautionary measures.

In general liability law will fill the 'gaps' of social security and private insurance; the employers' civil liability will generally be limited to the part of the victims damages which are left uncompensated. After all, most of the damages involved will be compensated by other, mostly insurance compensation schemes. As in English law, social insurance carriers have an action for reimbursement *vis-à-vis* the employer or colleagues of the victim, but this is limited to situations where the damage was caused by their intent or gross negligence. The reason for this is that both employers and employees are contributors to the social insurance system. According to the current view (and the same view was underlying the OW 1921) this must relieve them from having to bear the normal risk of civil liability. The general idea is that employers and fellow-employees should not pay twice for the same risk: both on an individual level, based on their civil liability and again, collectively, to finance social insurance benefits. This argument is particularly convincing for employers since they are the main contributors to the system.[159] Another argument for limiting the possibility of reimbursement claims despite its preventive effects is that disputes or litigation could cause agony and conflicts at the workplace. These political objections towards having to pay twice, as well as the desire to maintain a

---

[158]   A.T. Bolt/J. Spier, *De uitdijende reikwijdte van de aansprakelijkheid uit onrechtmatige daad* (1996), 92–93; S.D. Lindenbergh, *Arbeidsongevallen en beroepsziekten*, Deventer: Kluwer, 2000, 28.

[159]   There are also strong arguments to support proposals to change the law in this respect, see E.F.D. Engelhard, Regres van sociale verzekeraars, in: M. Faure/T. Hartlief, *Schade door arbeidsongevallen en nieuwe beroepsziekten* (2001), 47–76 (at 55) *et seq.* and E.F.D. Engelhard (*infra* fn. 187), 359 *et seq.*

good working atmosphere, explain why the employer does not generally have to face reimbursement claims.

6.3.3 Deviating Position of Dutch Law

It seems clear that in the Netherlands a refined system of social insurance is 88 used to carry and spread the damages caused by workplace accidents. As employers must make relatively high and many contributions to the funds of social insurance, they are the main contributors to the system. In this way it slightly resembles the industrial accident insurance schemes of the other law systems under review but unlike these other systems Dutch law ignores the cause of the personal damages it covers. Likely for this reason, the Netherlands showed a minor backlog with regard to readily acknowledging and diagnosing occupational diseases, as Philipsen points out in his contribution.

The fact that the damage was caused by a workplace accident, its 'risque professionnel character', is in Dutch law only of relevance if the victim brings in an action for damages (whereas in German and Belgian law civil litigation is excluded). As reimbursement actions by the social insurance carrier against the employer or colleagues of the victim are limited to intent or gross negligence, the risque professionnel character will only be exposed to the extent of the damages that were left uninsured.

6.3.4 Reintroduction of a Preferential Treatment of the 'Risque Professionnel'?

In 1989 it was estimated that for a mere 30 to 40 percent of disabled employ- 89 ees relying on social security the cause of their disability could not at all be related to their work.[160] Still, the scheme for pensions in case of long-term disability as discussed above makes employers bear the consequential risk of income loss exclusively, regardless of whether the occupational disability was caused by a workplace injury or otherwise.[161] Nevertheless over the years there have been strong voices favouring the re-introduction of special treatment for those employees whose damages were caused by workplace accidents. In 1992 the *Nederlands Instituut voor Arbeidsomstandigheden* (Dutch Institute for Labour Conditions, NIA) and the *Sociale Verzekeringsraad* (Social Insurance Council) organized a conference on the matter for professionals in the field, under the title 'Risque professional: The Netherlands being the outsider'. Although the 'risque social' principal is not fundamentally criticized (Fluit & Wilthagen 2001, 107), there have been voices to improve or at least separate basic protection for workplace victims. More recently former member of the conservatism party VVD (and since 2 September 2004 independent Member of Parliament) Wilders, who had raised his own more radical right wing political party, had taken the point of view that the WAO should ex-

---

[160] P.S. Fluit/A.C.J.M. Wilthagen (*supra* fn. 82), 107.
[161] References as to the – many – law reviews on latest developments of the *Occupational Disability Act (WAO)* reform plans are given by the SER http://www.ser.nl/scriptie/default.asp?desc=scriptie_sociaal_WAO_2.

clusively cover the 'risque professionnel'. Damages which are not caused at work should be left to a less protective scheme of social insurance.[162] This view was supported by some academics and the employers' lobby and gained force after the employers were made fully responsible for paying the WAO contributions and these contributions were made dependent on the disability rate of their employees (Rauws 2001, p. 109). This was perceived as inefficient and even unfair as the disability rate (and subsequently the employers' contributions) not only rises if employees suffer from an accident at the workplace but also due to accidents in the personal sphere.

What is more, in the midst of the plans involving the replacement of the WAO regime by the currently new IVA/WGA system, the Minister of Social Affairs planned to use the opportunity to introduce a special, more favourable insurance scheme for employees whose disability was caused by the 'risque professional', accidents or illness caused at work. The first draft to a new legislative proposal had been initiated by his Ministry in 2003. It introduced an improved regime for long term disability, that is occupational disability lasting more than two years, the so-called *Extra Garantieregeling Beroepsrisico's* (Additional Guarantee Provision for Occupational Risks, EGB).[163] This special 'risque professionnel' scheme would be made compulsory for all employers and would cover all employees who have been disabled to work due to occupational injuries or disease, providing them with full compensation for medical expenses and in the case of death the loss of life support.[164] According to the Dutch Minister this ruling followed from the International Labour Organization (ILO) agreements, by which the government was bound to offer permanent compensation to victims of occupational disability caused by workplace accidents.

90   Most recently however, it seems the Minister of Social Affairs has changed his plans, the Advisory Committee on Legislation had serious concerns on how the new social security regime of compensation for long-term disablement would relate to general social insurance, civil liability and the amount of claims. As a result the government has decided to wait; unclear is whether in the near future this newly planned scheme will be politically attainable, given the employers' (mainly small employers such as tradesmen) complaints that their financial burdens as aforementioned are already quite substantial.

---

[162] TK 2000–01, 27 402, no. 2 *(Sociale Nota)*, p. 133.
[163] Brief van Minister van Sociale Zaken en Werkgelegenheid, 16 september 2003, TK 2003–04, 28 333, no. 2.
[164] At Leiden University recently a proposal to a PHD-research project was initiated as to the possibility and desirability of this protection plan for employees, http://www.niwi.knaw.nl/nl/oi/nod/onderzoek/OND1300737/. The English translation of the new scheme's name was taken from this proposal.

6.4 The Current German Compensation Scheme (1970–2004)

6.4.1 Social Security Reform

Germany too has faced the need of cost containment in the 1970 and 1980 and   91
this, together with the desire to uniform different social security schemes and
changes in medical practice led to several changes. But the present social insurance system in Germany still protects against illness (*Gesetzliche Krankenversicherung*), incapacity to work (Rentenversicherung), and personal damages as a result of accidents (based on the basic mandatory *Unfallversicherung*,
UV). Next to social security law private law *Unfallversicherungen* are wide-
spread. The basic mandatory *Unfallversicherung* (Industrial Accidents Insurance, UV) was embedded in the federal *Sozialgesetzbuch* (SGB), which was
first introduced in 1975 and which forms the foundation for the current social
insurance system. Hereafter I will refer to it as the basic mandatory *Unfallversicherung* and what I mean by that is the aforementioned statutory accidents
insurance that is laid down in *section VII of the Sozialgesetzbuch* (Social Security Act, hereafter: SGB VII).

It covers in principle all employees, except for a few special categories
(e.g. civil servants, judges, soldiers) or when they are engaged in minor employment and the remuneration is above the annually assessed income ceiling
(€ 3,825 in West Germany in 2003). Similar to its neighbour countries though,
in Germany too, the mid-1970s 'mark the end of the so-called 'golden era' of
welfare state capitalism'.[165] Its social security schemes were, and still are,
marked with the need for cuts on their budget and social reform. In 1975 the
expansion of the welfare state came to an end and the first social security cuts
were put into force and throughout the 1980s the federal government's social
policy over-all aimed at consolidation and a decrease of the social insurance
contributions. The system covered approximately two thirds of the population.[166] Throughout the 1980s the increased social insurance contributions, for
one, were found to negative the competitiveness of German companies on an
international level. The federal government's policy focused on cost containment, until the unification between East and West Germany in 1990[167].

Special preventive measures were taken in social security law too, such as
the *Betriebliche Gesundheitsberichte* which the medical aid funds sent to employers informing them about diagnoses and new treatment programs and en-
couraging them to take action (see also the contribution by Philipsen). That is
hardly surprising as from 1990 onwards data on social spending show an increase larger than ever before, the Western German welfare system being extended to the former DDR, and reached its apotheosis in 1996. As known, the
former Kohl government's policy has been to use social insurance contributions to finance these costs and this has put employers under enormous pressure trying to compensate this extra expense by an increase of production. In

---

[165] M. Seeleib-Kaiser (*supra* fn. 121), 143.
[166] H. Kötz/G. Wagner, *Deliktsrecht* (2001), 581.
[167] In this sense M. Seeleib-Kaiser (*supra* fn. 121), 146 (with figures).

1997 the employers' contributions to social insurance schemes reached their highest point.[168]

92   What is the level of protection that in general, the basic mandatory *Unfallver-sicherung* left aside, has remained to exist for employees who are injured outside the scope of their employment? As for their need for medical care, Part V of the *Sozialgesetzbuch* (SGB V) offers full prepaid health care, albeit the major reform *Modernisierungsgesetz 2004* has reduced benefits considerably (while awarding bonuses for prevention). This so-called *Gesetzliche Kranken-versicherung* (Statutory Health Insurance, GKV) also provides for home care, household help, adjunct therapies, home health equipment and alike. It consists of more than three hundred funds that are governed by representatives of both employers and employees and it is said to cover 90 percent of the population, the remaining 10 percent are privately insured. All in all, aside from a few individual contributions for specific needs, all medical costs are covered by Federal and *Länder* Insurances. Special entitlements may be derived from local sickness funds, company sickness funds, trade funds, and licensed mutual health insurance schemes.[169]

Regarding their loss of income, employees are entitled to a continuation of wage payments by the employer for the first six weeks of absence (short term disability).[170] This is followed by social security's sickness benefits, based on the *Gesetzliche Krankenversicherung* (Statutory Health Insurance, GKV). Employees whose earnings exceed the ceiling must take out private insurance for sick-pay pensions (and medical costs). Below the ceiling an income related pension is paid for by the *Krankenkasse*, to the amount of 70 percent of the victim's gross income (not tax deducted) and with a maximum of 90 percent of his last earned net income, for a maximum of seventy-eight weeks. Due to the rather slow economic growth and the effects of the tax reform in 2000 this system has been under great pressure.[171] Nevertheless the risk of longer lasting income damage is still covered by social insurance, as well as additional costs of living as a disabled person, the costs of rehabilitation and the costs of nursing. If the disability resulted from an accident in the private sphere (or from a chronic illness) employees who have paid the contributions to the Statutory Pension Insurance for more than five years will be granted a disability pension. The amount of this depends on previous earnings and on a national standard period of employment. If the victim is partially able to work, a partial pension is granted.

---

[168] *I.e.* their contributions expressed as a percentage of gross wages. This information on data and figures are taken from M. Seeleib-Kaiser (*supra* fn. 121), 148.

[169] In 2003 about 90 percent of the population was insured through one of the more than 400 of these independent statutory health insurance companies, as claims M. Seeleib-Kaiser (*supra* fn. 121), 145.

[170] Par. 3 I Entgeltfortzahlungsgesetz. The Federal Red-Green coalition however did bring about a few major changes in this respect. Due to these changes skilled workers who are unable to continue to work in their profession no longer enjoy special treatment. They will have to either start a new career or rely on the regular disability program. See M. Seeleib-Kaiser (*supra* fn. 121), 153.

[171] See *e.g. Frankfurter Allgemeine*, 31 August 2004.

Special social security regimes apply such as the Railways Insurance, the Federal Social Insurance/sickness funds for miners, the Compulsory Health Insurance Scheme for Seamen, and the Agricultural Health Insurance Scheme. These are separate systems for invalidity, sickness and other medical expenses of railway workers, miners, farmers, seamen and artists. The special social security regimes are carried out by the Railways Insurance Institute, the Miners' Health Insurance Institution, or other industrial and agricultural accident insurance *Berufsgenossenschaften* (BGs).

How is all this for employees who suffer damages from workplace accidents?

6.4.2 UV Coverage

For employees who suffer from workplace accidents or occupational diseases 93 social protection is the strongest. For them, the Sozialgesetzbuch (Social Security Act, SGB) contains a special part, Part VII, as the mandatory *Unfallversicherung* or Industrial Accidents Insurance Act (UV) was integrated in this federal act in 1997,[172] due to which it has become even more a comprehensive system of social security insurance.[173] This industrial accident insurance still is their exclusive mandatory source of compensation, liability claims vis-à-vis third parties such as manufacturers left aside, and as said this has come to include the *unechte Unfallversicherung* throughout the years (see no. 64 above). There were some attempts to put a stop to the rising costs. Employers were expected to take more preventive measures in order to reduce the number of industrial accidents, for instance rehabilitation benefits were included, see above (no. 65). Subsequently in 1976 the employers' payments were made dependant not only on the risk factor regarding their branch of industry but also dependent on the number of accidents in their individual company.[174] Reputedly, this has decreased the number of work accidents in a significant way albeit at the beginning of the eighties the data on social spending show quite a rise.

Complaints have arisen as to the level of compensation offered by the *Unfallversicherung* of the *Sozialgesetzbuch*, section VII (SGB VII) or Industrial 94 Accidents Insurance Act (UV), although from a comparative point of view it generally seems to be reasonable (see below). It protects workers against the risk of bodily injury amounting to physical or mental impairment: Injured employees are entitled to medical health care (no money but rather care), rehabil-

---

[172] As said, originally the *Unfallversicherung* came into force by Law of 6 July 1884. The provisions of Book VII of the *Sozialgesetzbuch* were more recently introduced, in 1997, and came to replace Book 3 of the 1911 *RVO*.

[173] Contributory negligence of the victim is not taken into account, save in cases of intentional wrongdoing on his behalf. Furthermore, the abstract assessment of the victim's earning capacity can turn out to be more favouring to the victim than liability law's concrete assessment would be (although the insurer, due to statutory limitations or maxima, generally will not pay full compensation).

[174] In this way it strongly resembles the Dutch (former) WAO premium system, see nos. 78–79 above.

itation and disability pensions (*Verletztengeld* and *Verletztenrente*). Most favourable, compared to regular social security are the disability pensions: These are oriented at previous earnings and they depend on the degree of the disability (with a minimum of 20 percent and a maximum of 65 percent of pre-accident wages), save the expenses made during the first eighteen days of injury (for which there is the statutory pension insurance scheme and private accident insurance, based on a contractual insurance policy). To a certain degree however, the insurance payments to which the victim employee is entitled will be standardized. The loss of the right arm for instance is, reputedly, currently taxed as a loss of seventy-five percent of the victim's earning capacity (see U. Magnus, 'Compensation for Personal Injuries in a Comparative Perspective', 2000 *Wasburn* Law Journal 39). Remarkably all benefits are granted even if the injured employee continues to work full-time, because they are considered to be a form of compensation. In the case of death the widow or widower and minor children are entitled to the direct payment of one months wages (*Hinterbliebenenrente*) and funeral expenses and future pensions limited to 30 or 40 percent; orphans will receive more (*Witwen-, Witwer-* und *Waisenrenten*). Other damages such as property damage and non-pecuniary damages are not covered by this mandatory insurance.

In order to be compensated the damage must be caused in the course of and be related to an insured event, which in principle means employment accident, an *Arbeitsunfall*. This cumulative criterion seeks to exclude private matters that are attended during and in the course of employment, such as buying groceries or seeing a friend. In § 8 of section VII of the *Sozialgesetzbuch* (Social Security Act, SGB), this is explained as an external temporary event which has caused the bodily injury or death of the employee. Due to the enlargement of the insurance coverage as was discussed above this now also includes the regular attendance at kindergarten, school and universities, the rescue attempt of a person in need, donation of blood and other socially desirable activities.

In all of these cases the employer (kindergarten, school, university, hospital and alike respectively) nor fellow employees may be faced with a liability claim for damages (§ 636 and 637 *Reichsversicherungsordnung*; cf. § 104 and 105 of section VII of the Sozialgesetzbuch, the SGB VII). An exception is still made, as I mentioned this before, for situations in which the damage was caused by the employers' or fellow employees' intent or gross negligence. See for further details below, no. 96.

95   The industrial accidents insurance (*Unfallversicherung*) of the *Sozialgesetzbuch*, section VII (SGB VII) is currently administered by ninety separate legal entities, mostly *Berufsgenossenschaften* (BGs), which each represent an industrial, agricultural and public service accident insurance fund, supervised by the Minister of Labour and Social Security. The system is financed entirely from contributions made by employers. Public officials are in principle by law obligated to be insured in the statutory social insurance, which means they are members of the statutory health, long-term care, industrial (and other) accidents, pensions and unemployment schemes. The public

employer and the official share the costs of social insurance in accordance with the applicable contribution rate which is a defined percentage of the gross income, but as said the costs of the statutory accident insurance are borne by the employer alone. The insurance system is administered by special insurance funds for public sector at Federal and *Länder* level, which are financed by revenues from taxes. Finally, the self employed must take out private health and disability insurance.

6.4.3 Civil Liability Law

As said under limited circumstances the victim-employee has the right to sue   96
his employer or a fellow employee for personal damages. Also, in German law all of the aforementioned social insurance carriers (*Renten-, Kranken-* and also our *Unfallversicherungsträger*) have an action for reimbursement vis-à-vis the liable party. General rules as to reimbursement actions of social insurance carriers are laid down in the *Sozialgesetzbuch* (SGB). But for liability actions by the victim-employee and for actions for reimbursement of his social insurance carrier (the *Berufsgenossenschaft* in case of workplace accidents), the *Reichsversicherungsordnung* (RVO) gives special rules.

Above I briefly touched upon the limited right of action of the victim-employee (students, blood donors and alike) towards his employer (university, hospitals and alike) in case of workplace accidents. Both the employer and fellow employees almost enjoy immunity to civil liability claims for damages. The employer or fellow employee may still be held liable by the victim if they caused the workplace injury intentionally (*vorsätzlich*) or if the injury was an *Arbeitsunfall* (see no. 94) though caused '*bei der Teilnahme am allgemeinen Verkehr*' (§ 636 RVO) and for property damage. However as far as personal damages are concerned the victim will only be entitled to compensation for those remaining damages that are not yet covered by insurance.

To the extent that the *Berufsgenossenschaft* (BG) has paid the victim, his employer or colleagues can only be held liable if they have caused the accident by personal intent or gross negligence (§ 640 RVO cf. § 110 SGB VII). For his intent or gross negligence the employer or fellow will be liable if the *Berufsgenossenschaft* (BG) seeks reimbursement for the damages that were paid to the employee based on the statutory insurance. Actions for reimbursement against the employer or fellow-employee require proof of gross negligence or intent and will therefore be rare. Moreover, different from Dutch law, the *Berufsgenossenschaft* (BG) has been granted a discretionary power as to the decision whether or not to seek reimbursement (§ 640 section 2 RVO). The employer and fellow employees thus enjoy an almost full immunity to civil liability (cf. § 104 and 105 SGB VII).[175] The reason for this is that the employer is considered to have already paid for the victim employees' damages indirectly, through the insurance contributions (which are exclusively paid

---

[175] *De facto* this means that the employee cannot recover damages for pain and suffering (as opposed to plaintiffs with a valid liability claim, see § 847 BGB). Some of the aforementioned leading authors claim this to be the main disadvantage of the German system, compared to regular fields of tort law, B.S. Markesinis/H. Unberath (*supra* fn. 17), 727).

for by employers). Likewise, the fellow employee, if faced with the risk of liability, would be able to seek reimbursement vis-à-vis the employer, which explains why he too is freed from the risk of negligence liability.

If a third party (*i.e.* other than the victim's employer or colleague at the
time of the event) has caused the damage, then the paying social insurer will
be subrogated into the victim's claim without restrictions (§ 116 SGB VII).
However, in the case of shared liability with the employer or colleague the
third party will only be subjected to an action for reimbursement amounting to
his share of the blame. Civil liability law is still regarded as a valuable supplement to the statutory insurance, or, as Magnus (op. cit. *supra*, no. 95) claims:

> 'The accident insurance scheme has proven to be an absolutely necessary
> and helpful addition to, and partial substitution for, civil tort law. It is
> sometimes criticized as providing insufficient compensation. However, this
> disadvantage might strengthen potential victims' efforts to avoid damage
> themselves. Additionally, the possible recourse action of the accident in-
> surance agency against the original tortfeasor may induce the latter to
> avoid damage whenever possible. Thus, tortuous liability still plays the
> background melody'.

6.5   The Current Belgian Compensation Scheme (1970–2004)

6.5.1   *Arbeidsongevallenwet 1971 (Industrial Accidents Act, AoW): From Personal to Collective Liability and Mandatory Insurance*

97   Although Belgian law resembles German law in the sense that industrial accidents are covered by a special industrial insurance, for decades both systems
differed by the fact that the Belgian no-fault insurance was made voluntary
and still based on the idea that it was the employer's personal obligation to pay
instead of a collective liability as in German law. In 1971 this came to an end.
The *Arbeidsongevallenwet 1971* (Industrial Accidents Act, AoW) has made
the industrial accidents insurance compulsory and it still is today, covering
about 2.5 million employees.[176] Under this statutory insurance scheme employers are now under the obligation to take out this insurance and to make the
insurance contributions. The fact that only employers pay is similar to German
law and different from England and the Netherlands.[177] It is based on the risk

---

[176] *Eeuwfeest van de Arbeidsongevallenwet en 50 jaar preventie, Het Belgische model: een voorbeeld met een eigen meerwaarde*, report published by the *Beroepsvereniging der Verzekerings-
ondernemingen* (BVVO), p. 3. Article 6 of the 1971 Act rules that the employment contract
may not be nullified with regard to the Act (subs. 1) and that any contract contrary to its rulings is invalid (subs. 2).

[177] The English National Insurance (insurance against the risk of income loss) is financed through
contributions shared by both employers and employees and through state subsidies by raising
taxes; the National Health Service (medical care) is completely State financed (from tax
money). In the Netherlands social insurance for the risk of invalidity is financed almost completely by employers' contributions and the state, comparatively spoken, hardly participates in
financing this and other (medical aid) social schemes for employees. In both law systems additionally voluntary liability insurance is taken out and paid by most employers.

theory (*'théorie du risque'*): 'the working class' needs protection against the risk of industrial injuries.[178]

Different from German and English law (as well as from the Belgian compensation scheme for occupational diseases, see below), the statutory insurance is carried out by – commercial – private law entities, listed in the *Arbeidsongevallenwet 1971* (Industrial Accidents Act, AoW) (art. 49). The state supervises the financial soundness and solvability of the insurance companies as well as watches over the rights of employees to be insured. In all instances the relation between the insurance company and the employer is based on the (private) insurance contract; the employer chooses the insurance company, which will give the insurance companies an incentive to compete. The *Commissie voor het Bank-, Financie- en Assurantiewezen* (Committee for Banking, Finance and Insurance) is responsible for financial supervision of the market. If the employer in violation of the *Arbeidsongevallenwet 1971* (Industrial Accidents Act, AoW) has not taken out the statutory insurance he will be taken to participate in – and accordingly pay contributions to – the special public law *Fonds voor Arbeidsongevallen* (Fund for Industrial Injuries), which will then take the victim-employee's request for compensation into consideration. This fund is governed by both employers and employees and paid for by the insurance companies and contributions of employers. The fund protects the insured persons' rights and entitlements under the *Arbeidsongevallenwet 1971* (Industrial Accidents Act, AoW) and in doing so it is also responsible for the technical and medical supervision.

In 1981 this industrial accident insurance was *formally* integrated in the social 98 security system (*by the Act of 29 June 1981*); the entitlement to insurance awards is since then seen as a basic right which is directly derived from public law's protection. All employees whose accident risks are covered by social insurance are covered by this workplace insurance scheme, that is: employees under a contract of service as well as trainees under an agreement of apprenticeship. Comparable to German law and different from the English IIS (which is more restrictive), temporary (summer job) employed students and season employed workers are also included. In order for the victim-employee or his relatives to receive benefits it must be shown that his injury was caused by an '*arbeidsongeval*' (workplace accident). The Belgian Supreme Court (*Hof van Cassatie*) has given a rather free interpretation to the requirement that the victim-employee's damages must be caused by an 'accident'. It will suffice that he shows a sudden event in his daily routine that may have caused his injury. Reputedly lower courts have been reluctant to use this wide interpretation (Rauws 2001, p. 115).

As far as the relation to his work is concerned the Belgian legislator defined the term as any accident suffered '*by and* in the course of employment' causing personal injury, including accidents on the way to and from work. If the victim-employee at the time of the event was virtually under the control of

---

[178] The same foundation was underlying the old Dutch OW, see W.J.P.M. Fase (*supra* fn. 116), 52.

his employer (*i.e.* when lacking personal freedom) the accident will be presumed to have occurred in the course of employment. Again similar to English law, the accident will be presumed to have been caused by the employment if it occurred during work. If the damages were caused by intentional acts or omissions of the victim-employee himself he is not entitled to insurance benefits. For accidents which occur in the victim-employee's private life a special insurance can be taken out by the employer on a voluntary basis.

Most procedures it seems are procedures for lower courts that are concerned with the calculation of the exact rate (percentage) of disability (Rauws 2001, p. 116). This statutory industrial accidents insurance offers compensation for temporary and permanent disability, either on physical or mental grounds, as well as wrongful death and it also intermediates in medical expenses. Interestingly, nowadays disability pensions seem relatively high: 90 percent of the average daily salary, with a statutory maximum. For employees whose income exceeds the statutory maxima special additional insurance coverage can be taken out by the employer on a voluntary basis. The main condition for the statutory insurance payments is that the worker had been exposed to risks, which is presumed to be satisfied when the employee worked in an enterprise that was listed by Royal Decree. Provided these conditions were met, the *Arbeidsongevallenwet 1971* (Industrial Accidents Act, AoW) generally[179] obliged insurers to compensate the income as said, albeit with a statutory ceiling. In the case of temporary disability medical and reintegration costs as well as the loss of income are protected; in the case of permanent the statutory insurance offers the right to compensation for the costs of maintenance and replacement of prostheses as well as disability pensions. These pensions will be fixed in cases where the disability rate is below sixteen percent and indexed if the victim is disabled by sixteen percent or more. Lastly, in the case of wrongful death certain relatives such as the employee's widow, ex-wife and children are entitled to an annuity for their loss of financial life support as well as to compensation for the funeral expenses. Property damage and non pecuniary (immaterial) damages are not included (and therefore probably in most cases not compensated, since the employee has very limited actions to civil liability).[180] The insurers of these risks however complain the workplace accidents insurance is hardly profitable as the transaction costs are high; reputedly the profits are almost exclusively made by investment profits ands not by earnings (Rauws 2001, p. 118). Compared to liability law's awards the level of compensation seems reasonable and the employers' contributions are said to be moderate. It must be seen though that these do not include the additional contributions that employers owe for the – separate – risk of occupational diseases.

---

[179] The act gives special rules for see- and sportsmen.
[180] See Rauws 2001, p. 113 and http://socialsecurity.fgov.be/.

6.5.2 Civil Liability Litigation

Additional protection is offered by civil liability law, for instance in cases   99
were the damage was caused by a third-party, the employer or colleagues of
the victim. Actions for damages by employees against their employer or colleagues however are exceptional because of special limitations in civil liability
law. Employers can only be held liable for property damage of the employee
and for intentional wrongs, accidents to and from work and other traffic accidents. Actions for damages against colleagues ('*aangestelden*', cf. art. 1384
subsection 3 Belgian *Burgerlijk Wetboek* (Civil Code, BW), that is: those who
have the same employer as the victim-employee are subject to the same limitations. In 2000 the legislator added a clause to article 46 of the *Arbeidsongevallenwet 1971* (Industrial Accidents Act, AoW) which now allows the victim to
also claim damages from his employer if the latter has seriously violated the
rules on wellbeing.[181] As far as the victim's damages fall under the industrial
accidents insurance, the insurer (or the fund if the employer was *not* insured)
will be subrogated into the victim's action for reimbursement vis-à-vis a third
party who can be held liable towards the injured employee and – in the exceptional situations aforementioned – even against the employer of the injured
worker.

7. Summary

7.1 General Remarks

In all four countries under review compensation for victims of workplace acci-   100
dents is faced by a combination of compensation systems: social security, State
benefits, private insurance and civil liability law. The variety in these combinations of solutions is great. In German law and Belgian law, and to some degree
in English law, special protection for workplace accidents is mainly offered *by
basic insurance* with strong public law elements such as a mandatory character, a state regulated minimum and/or maximum level of compensation and
state controlled administration. On the contrary, in Dutch law special protection cannot be found in basic insurance (as for social and private insurance the
cause of the damage is irrelevant which means personal and work injuries are
treated equally). Here, special protection for workplace accidents is mainly offered *by additional liability claims* (claims in addition to the basic insurance
compensation which does not offer any special protection) and it has, to that
extent, a mere private law character (no mandatory insurance). However, the
requirements for the right to compensation are not far apart as Dutch liability
law *de facto*, compared to the three other countries, has a far reaching liability
for employers. Rather the rules of special protection offered by each of the
four countries show great differences with regard to the scope of protection
(who benefits from this special protection), the financial resources (who is
paying) and the systems' effectiveness (operational and procedural aspects,
costs and deterrents).

---

[181] Article 46 subs. 1, sent. 7 *Arbeidsongevallenwet 1971*.

The workplace accidents insurances ruled by German law and Belgian law are paid for exclusively by contributions of the employer, while in current Dutch and English law the risk of workplace accidents is borne by contributions of employers and their employees (and to some degree the state). The level of protection offered by (the concurrence of) these law systems show great variety too. One century ago, in 1906, Pelster claimed that law systems will *at the very minimum*[182], allow the injured employee or his financial dependants to recover from the employer, a colleague or any other party *who due to his personal fault* caused the damage, as a principle of civil liability. *At the very most*, he argued, the victim or his dependants will be entitled to a full state pension or allowance (without further ado and regardless of issues concerning civil liability)[183]. In today's legal complexity it seems more common than it is exceptional, to be dealing with at least several concurrent compensation schemes for workplace accidents per national law system. These systems may each serve justice and/or help to prevent the occurrence of accidents (or of an increase of damages), or have other purposes.[184] As Philipsen shows in his contribution, special social security or labour law rules as well as the threat of civil liability actions may serve as incentives for employers and employees to prevent the number of workplace accidents or the amount of damages from increasing.[185]

101   Civil liability law, social and private insurance work highly interactively. Decisions with regard to the scope of protection of one compensation system may have important ramifications for other systems. This is hardly surprising. To a great extent tort law, social insurance and private law are in fact offering the same thing: compensation for income losses, medical health costs and loss of life support and incentives as to the occurrence of these damages. But civil liability law, for one, has serious barriers for the right to full recovery such as the fault requirement, the requirement of recoverable damages and the concept of contributory negligence. The more difficult it will be to obtain compensation this way, the more need there will be for alternative ways of compensation, for instance based on private insurance. The financial risk of civil liability also seems less spread, especially if liability insurance is not mandatory than it is with more general insurances, such as basic accidents insurance or social insurance. This could be a reason to not award damages that are already covered by social security or private insurance, instead of vice versa. So too, a decrease in the accessibility of private or social insurance (*e.g.* due to its rising cost) may result in a growing need for tort law's protection. Given the fact that

---

[182] H.J.W. Pelster, *Ongevallenverzekering. Eene Inleiding tot de Studie van Wetten betreffende de Schadevergoeding aan Werklieden ter zake van Bedrijfsongevallen* (1906), 86.

[183] H.J.W. Pelster (*supra* fn. 182), 87.

[184] See the contribution of Hoop, in this book.

[185] See the contribution of Philipsen. I assume that as far as civil law actions may serve as a deterrent with regard to workplace accidents, they will modestly add up to the various public, social and labour law rules and conditions imposed on employers, demanding a safe and accommodating workplace.

public financial resources are limited, special rules may then seek to prevent victims from being compensated more than once.

At the end of the day each system will be directed by its respective end-goals. 102 Social security law seeks to guarantee or maintain a certain level of subsistence, based on the concept of solidarity,[186] while tort law's main goal is to promote corrective justice.[187] Policy choices with regard to the level of protection offered by each system may, in other words, explicitly or implicitly be directed by different end-goals.[188] As a result social security cuts, while based on certain policy considerations, may have unwanted ramifications for tort liability (for instance cause more claims which might even result in higher liability insurance premiums) or private insurance (for instance it may create new private insurance markets and/or with an analogous decrease in the need for other insurances).

7.2 Basic Findings

In order to grasp such ramifications one needs to know the ways in which (and 103 the reasons why) the workers cause is addressed in the law systems under review. More so, the developments and changes in earlier years will give us a better understanding of these processes and how they work. Political aspects such as social disarray or dissatisfaction of employees as to their legal position (poor work conditions, poor wages, high medical expenses, high contributions, dependency of their employer) and the labour unions' role in the workers' cause seem to have asked for changes of the law. Let me briefly summarize the basic findings: to what extent were those solutions taken to be

---

[186] S. Klosse, *Menselijke schade: vergoeden of herstellen? De werking van (re)integratieregelingen voor gehandicapten in de Bondsrepubliek Duitsland en Nederland* (1989), 263; K.P. Goudswaard/C.A. de Kam/C.G.M. Sterks, *Sociale zekerheid: op het breukvlak van twee eeuwen* (2000), 202. As known the concept of solidarity may be used in various ways, highly normative (in connection with social movements against injustice) or more neutral (referring to people's interdependency, either on a philosophical or on a practical, occupational level). K. Tinga/E. Verbraak, Solidarity: An indispensable concept in social security, in: J.P.A. van Vught/J.M. Peet (eds.), *Social Security and Solidarity in the European Union. Facts, Evaluations, and Perspective* (2000), 254–269 refer to social security as 'also an ethical issue, reflecting both the self-interest of the well-to-do *vis-à-vis* the old, the sick, the unemployed, and the poor' (p. 254). To them these issues fully reflect the concept of solidarity and in that very sense, as the backbone of social cohesion it must be understood here too. See further P.S. Fluit, *Verzekeringen van solidariteit* (2001), 1.

[187] Surely the concept of (distributive) justice plays an important role in social security law too, and not just on a philosophical level but also in practice. Schokkaert argues that it should play a dominant role in today's social reforms (in: E. Schokkaert (ed.), *Ethics and Social Security Reform* (2001), 3–35). Tort law's goals too have been and still are highly debated and object of a (polemic) controversy. See more recently in Dutch doctrine a.o. T. Hartlief, *Ieder draagt zijn eigen schade* (1997), M. Faure/T. Hartlief, *Nieuwe risico's en vragen van aansprakelijkheid en verzekering* (2002), 8 *et seq.* and my thesis E.F.D. Engelhard, *Regres* (2003), 199–228 (with further references on further literature and empirical studies).

[188] As will be seen below these fundamental differences have been blurred over the years, as new, economical, private market ideas have set foot in social security law (and as tort law came to be more concerned, or so it seems, with reallocating losses more equally and fairly).

satisfactory and what ramifications did earlier choices have for new solutions or improvements? How did each of these processes evolve?

104   In nineteenth century Germany the biggest hardship for employees seeking compensation for industrial accidents was the proof that the employer or other workers had been at fault. The *Reichshaftpflichtgesetz 1871* (Liability Act, RHpflG) was an improvement in this respect as it reversed the burden of proof for some industries and introduced vicarious liability for others. In the following years however the right to compensation of employees was hindered by remaining fault elements (proof that fault had been missing for instance by an Act of God defence and contributory negligence), the fact that employers – even if found liable – would not always be insured and therefore insolvent and the risk of workplace conflicts which was inherent to civil liability litigation.

The German solution was found in the mandatory *Unfallversicherung* (Industrial Accidents Insurance Act, UV), a statutory insurance which was made compulsory (and with a mandatory coverage), based on limited and fixed awards. This insurance system came under pressure in the first half of the twentieth century, when the awareness of the need for protection (the ideal of welfare state) had spread to non-industrial accidents and other causes of personal detriment.

This has led to a widening of the protective scope of the *Unfallversicherung* (Industrial Accidents Insurance Act, UV), which seems now more related to the employee in person and his professional relationship with the employer than with the actual work and work hazards, albeit with moderate benefits (*e.g.* disability pensions limited to a maximum of 66 percent of the victim's last wage, limited medical expenses and alike).

105   As for English law, already in the nineteenth century the right to compensation based on liability law of injured employees was vested in torts that generally required fault. Vicarious liability was to no avail for employees as the employer would escape liability with the defence of common employment. The 1880 Employers' Liability Act served to eliminate this unequal position of employees compared to non workers, who could invoke the principle of vicarious liability. Still the position of employees was hindered by the requirement of fault and the fact that the 1880 Act had very limited (fixed) awards as well as by the adversarial nature of liability law and the risk of workplace conflicts. Employers were unsatisfied too as they feared more claims and so they raised Friendly Societies, private funds that would compensate but would make the employee give up their action in liability law. These were found problematic as they barred recovery (while the employee was in no bargaining position to refuse to give up his right) as well as the fact that the Friendly Societies gave no incentives to take care (deterrence).

The Workman's Compensation Act 1897 (WCA) introduced a regime of strict liability, which improved the employees' rights in liability law. Still the system was not satisfactory, for two (main) reasons. First, this system still had a strong adversarial nature: It forced the employee to claim damages from his employer personally as it did not require the employer to be insured (which meant one also risked insolvency). Furthermore, its level of compensation was reduced to only 50 percent of the victim's damages while the latter still risked being bound by contracting out agreements.

Relief was found in both ways with the abolition of the Workman's Compensation Act 1897 (WCA) and its replacement by social security law as the primary source of compensation. As for their medical costs, injured employees were treated no different from other accident victims. For their loss of income, a special system was introduced by the National Insurance (Industrial Injuries) Act 1946. This entitled workplace victims to 'full' disability pensions in the sense that the 50 percent reduction of damages was replaced by compensatory awards, albeit still moderate pensions. Additional protection was offered by common law rules on liability. As the level of social security compensation was still moderate, liability law was improved in at least two ways. Firstly, establishing liability was made easier (as concrete statutory duties of care were imposed on employers, fault was made easier to prove and vicarious liability was made possible by the prohibition on the defence of common employment). Secondly, the Employers' Liability (Compulsory Insurance) Act 1969 introduced mandatory liability insurance.

The more preferential treatment of employees who suffered from workplace accidents or disability has been furthered by the Social Security (Recovery of Benefits) Act 1997. This special legislation arranges that – seemingly – all social security benefits to employees are re-paid by any person who can be held liable for the victim's damages, including his employer.

In Belgian law the compensation for employees was originally to a certain extent found in the so-called Financial Aid and Prevention Fund 1886, but this was perceived as too limited. Civil liability law (art. 1382 and 1384 of the Belgian Civil Code) generally centred on fault, which was thought to be a serious barrier to compensation. 106

As a result the Arbeidsongevallenwet 1903 (Industrial Injuries Act, AoW) introduced a regime of strict liability: The employer would be liable for the damages of employees who suffered from a workplace injury. This was not satisfactory for at least three reasons: The system was limited to physical injuries, more importantly the compensation that was offered was limited to fifty percent of the victim's damages and last but not least the system rested on the individual shoulders of the employer. The employer could take out (private) liability insurance but that was voluntary and the victim's right to compensation was in principle vested in his personal, civil law obligation against his employer – not vis-à-vis the insurer of the latter.

The amount of compensation was improved in 1930: The compensation of income loss was raised from fifty percent to two thirds but this did not apply for the first month of income loss. In 1951 the complaint to the limited compensation was more properly addressed as the first months awards were raised to eighty percent of the victim's income and thereafter the fully disabled employee would receive ninety percent. Similar to German law the scope of protection was widened, in 1951 for instance as the system came to cover commuting accidents.

A more drastic change came about in 1971 as the system came to rely on the principle of insurance. The Arbeidsongevallenwet 1971 (Industrial Injuries Act, AoW) introduced a system of mandatory accidents insurance similar to German law, which entitled victims of workplace accidents to the insured sum and was based on contributions by their employer. In 1981 the system was made part of social security, as in German law, though with a more limited scope of protection but with relatively high disability awards. Actions for damages by employees against their employer or colleagues are exceptional because of special limitations in civil liability law. The Belgian insurance generally seems to work quite to the satisfaction of employers and employees. Although the level of compensation is limited in the sense that property and non pecuniary damages are excluded and that personal damages are fixed, it is relatively high compared to the level of compensation in liability law for the kinds of damages that do fall within the scope of insurance and the employers' contributions seem to be moderate (although these do not include the additional contributions that employers owe for the – separate – risk of occupational diseases). The insurers of these risks however complain the workplace accidents insurance is hardly profitable.

107   Similar to the first three law systems originally the workers cause in Dutch law was addressed by fault based liability. Here too, liability law's adversarial nature (with the risk of workplace conflicts) as well as the proof of fault and other fault related elements (the contributory negligence defence) were serious barriers for the employees' right to compensation. The introduction of the Ongevallenwet 1901 (Industrial Injuries Act, OW) put an end to both barriers, as the employees' right to compensation for their medical expenses and income loss was now made independent of fault while it was the insurer – not the employer personally – who carried these risks. But as the protection against basic risks was more and more seen as a basic right (and the scope of the system gradually widened) the Ongevallenwetten 1901 and 1921 (Industrial Injuries Act, OW) heaped criticism: Why should industrial accidents have to be treated any different from accidents that occur outside working hours?

Finally in 1967 this system was replaced by systems of social insurance, covering the basic risks of medical expenses and income loss regardless of the way they were caused. As a result the current Dutch position of employees who suffer personal damages caused by workplace accidents is from the outset quite different from German, Belgian and English law. The employer is not subjected to mandatory occupational insurance. In Dutch law he faces the – much wider – financial risk of disability *regardless of its cause* as he is held to continue his employees' salary payments during the first two years of work related disability as well as to most of the social security premiums. It must be noted though that many employers take out additional private insurance for their employees, which is limited to the risk of workplace accidents, although on a voluntary basis. Additionally, the employer does face the risk of civil liability for accidents caused by violations of safety rules or of good employment (contrasting German and Belgian law, where the sanction to such violations is not primarily that he risks liability, provided they were not committed intentionally, but that his no-fault accidents insurance premiums will be raised). In the more exceptional cases of gross negligence or intent by the employer he is charged with the obligation to repay the victim's benefits *vis-à-vis* the social insurance carrier.

More recently social security cuts were introduced by which employees who have been partly disabled were strongly limited in their right to sick pay. It seems likely this may result in an increasing demand for private disability insurance (and possibly in a growing need for liability insurance coverage), to be taken out by the employer. There are even voices to re-introduce a special no-fault insurance for the 'risque professionel' similar to Belgian or to German law for why, it is asked, should the employer have to pay for personal risks run by his employees in their private lives such as going on a skiing trip? This argument gained force after the employers were made fully responsible for the contributions to disability pension schemes and their contributions were made dependent on the disability rate of their employees (which includes accidents in the personal sphere that are beyond the employer's control). But these voices that were raised in Parliamentary debates and literature are not likely to be followed in the near future.

7.3 In Fine

As was shown, for employers facing the cost of workplace accidents the true meaning of these differences between the respective law systems must be judged by analyzing the collateral source rules, rules of priority and actions for reimbursement in (national) *liability* law. My analysis shows that in the end these rules will decide who should be the final paying party. In English law for instance, social security provides for special earnings related benefits for workplace victims, but based on special reimbursement rules it is the employer who will ultimately have to re-pay these amounts to the Secretary of General. That is, if held liable for the event. Surely then, the more strict liability rules are, the more costs will be re-allocated in this way from the community (taxpayer) back to the *employer* (be it that the latter may charge the risk of having to bear these costs to the consumer).

108

In Germany, Belgium and the Netherlands however, the insurer who has compensated the injured employee has a very limited right to receive reimbursement from the employer; the latter is – aside from exceptional grounds such as gross negligence or intentional wrongdoing – freed from the risk of such claims. This seems logical for Belgian and German law as the employer – and only the employer – directly pays for the workplace insurance (leaving aside that from a more fundamental point of view one might be critical towards this 'immunity': Are the reasons for paying these contributions really the same as the reasons for facing liability?). In this respect one could say that while English law makes the employer pay through his part of the social security (workplace related) contributions completed with his contributions for third party liability insurance, German and Belgian law do so by making him pay for the workplace insurance. In all three the costs will then ultimately be borne by the employer, be it that in German and Belgian law the financial risk he is carrying seems bigger (viz. not dependent on grounds for liability).

Dutch law again seems different (no obligation to contribute to a special insurance for workplace injuries, nor the risk of reimbursement claims by the insurers who carry these costs). As was shown however, these days the employer in the Netherlands is already bearing the (bigger) risk of social insurance benefits exclusively as well as the risk of liability claims based on the very stringent Dutch regime for strict liability. In that respect it seems at the very least likely that both him, and the employee (who is given the right to sue his employer for additional earnings related damages) would be better off if Dutch law tried on its neighbour countries coat.

Bibliographic References

AARTSEN, L. & DE JONG, PH. (eds.), *Op zoek naar nieuwe collectiviteiten. Sociale zekerheid tussen prikkels en solidariteit*, Den Haag: Elsevier, 1999.

ABRAHAM, K.S., Twenty-First Century Insurance and Loss Distribution, in: *Tort Law, Public Law and Legal Theory*, Working Paper Series, Paper 1, 2004 (published at http://law.bepress.com/uvalwps/uva_publiclaw/art1).

ABRAHAM, K.S., Liability Insurance and Accident Prevention. The Evolution of an Idea, *Public Law and Legal Theory*, Working Paper Series, Paper 2, 2004 (published at http://law.bepress.com/uvalwps/uva_publiclaw/art2).

ADRIAANSENS, H.P.M. & ZIJDERVELD, A.C., *Vrijwillig initiatief in de Verzorgingsstaat, cultuursociologische analyse van een beleidsprobleem*, Deventer: Van Loghum Slaterus, 1981.

BARENTSEN, B., *Arbeidsongeschiktheid. Aansprakelijkheid, bescherming en compensatie*, Deventer: Kluwer 2003.

BARTA, H., *Kausalität im Sozialrecht. Entstehung und Funktion der sogenannten Theorie der wesentlichen Bedingung*, Berlin: Duncker & Humblot, 1983.

BEVERIDGE REPORT, *Social Insurance and Allied Services*, Report by Sir William Beveridge, Command Paper 6404, London: Her Majesty's Stationary Office, 1942.

BOLT, A.T. & SPIER, J., *De uitdijende reikwijdte van de aansprakelijkheid uit onrechtmatige daad*, Zwolle: Tjeenk Willink, 1996.

BERGHAHN, V., Growth, industrialization and social change, in: Breuilly, J. (ed.), *Nineteenth Century Germany. Politics, Culture and Society 1780–1918*, London: Arnold, 2001.

BIER, L., *Aansprakelijkheid voor bedrijfsongevallen en beroepsziekten*, Deventer: Kluwer 1988.

BONA, M. & MEAD, PH. (eds.), *Personal Injury Compensation in Europe*, Deventer: Kluwer, 2003.

BRUNSCH, D., Employer Services, in: Guidotti, T.L. & Cowell, J.W.E. (eds.), *Workers' Compensation, special edition of Occupational Medicine*, vol. 13, no. 2, 1998, pp. 345–355.

CANE, P., *Atiyah's Accidents, Compensation and the Law*, London/Edinburgh/Dublin: Butterworths, 1999.

COHEN, P., *The British System of Social Insurance. A History and Description of Health Insurance, Widow's and Orphan's Pensions, Old Age Pensions (Contributory and Non-Contributory), Unemployment Insurance, Workmen's Compensation and Industrial Assurance*, London: Philip Allan, 1932.

CARR, J.D., Workers' Compensation Systems: Purpose and Mandate, in: Guidotti, T.L. & Cowell, J.W.E. (eds.), Workers' Compensation, special edition of *Occupational Medicine*, vol. 13, no. 2, 1998, pp. 417–422.

DEAKIN, S. & MORRIS, G.S., *Labour law*, London a.o.: Butterworths, 1998.
ENGELHARD, E.F.D., Regres van sociale verzekeraars, in: Faure, M. & Hartlief, T. (eds.), *Schade door arbeidsongevallen en nieuwe beroepsziekten*, Den Haag: Boom, 2001, pp. 47–76.
ENGELHARD, E.F.D., *Regres*, Deventer: Kluwer, 2003.
FASE, W.J.P.M., De legitimering van het verplichtend karakter van de sociale verzekering, in: Jaspers, A.Ph.C.M. a.o. (eds.), *'De gemeenschap is aansprakelijk...' Honderd jaar sociale verzekering 1901–2001*, Lelystad: Koninklijke Vermande, 2001.
FAURE, M. & HARTLIEF, T. (eds.), *Verzekering en de groeiende aansprakelijkheidslast*, Deventer: Kluwer, 1995.
FAURE, M. & HARTLIEF, T., *Nieuwe risico's en vragen van aansprakelijkheid en verzekering*, Deventer: Kluwer, 2002.
FLUIT, P.S., *Verzekeringen van solidariteit*, Deventer: Kluwer, 2001.
FLUIT, P.S. & WILTHAGEN, A.C.J.M., Het risque social, in: Jaspers, A.Ph.C.M. a.o. (eds.), *'De gemeenschap is aansprakelijk'... Honderd jaar sociale verzekering 1901–2001*, Lelystad: Koninklijke Vermande, 2001, pp. 107–125.
FONTAINE, M., La réforme sur les accidents de travail en Belgique et la distinction entre assurances de choses et assurances de personnes, in: Bedour, J. *et al.* (eds.), *Études offertes à André Besson*, Paris: Librairie générale de droit et de jurisprudence, 1976.
FREDERICQ, S., *Moderne risico's en vergoeding van letselschade*, Brussel: Bruylant, 1990.
DE GIER, H.G., VAN WIJNGAARDEN, P.J. & ROELOFS, A.M.E., *Sociale zekerheid in Europa: trends en perspectieven*, Utrecht: LEMMA, 1994.
GOUDSWAARD, K.P., DE KAM, C.A. & STERKS, C.G.M., *Sociale zekerheid: op het breukvlak van twee eeuwen*, Alphen aan de Rijn: Samsom en Deventer: Kluwer, 2000.
GUIDOTTI, T.L., Effective Intervention to Reduce Occupational Injuries in Alberta: A Case Study of Financial Incentives, in: Guidotti, T.L. & Cowell, J.W.E. (eds.), Workers' Compensation, special edition of *Occupational Medicine*, vol. 13, no. 2, 1998, pp. 443–449.
HANES, D.G., *The First British Workmen's Compensation Act 1897*, New Haven/London: Yale University Press, 1968.
HARTLIEF, T., *Ieder draagt zijn eigen schade*, Deventer: Kluwer, 1997.
HERMANS, P.C. & PRINS, R. (eds.), *Causaliteit en arbeidsongeschiktheid. Verslag van een studiedag over het 'risque professionel' in Nederland*, uitgave Sociale Verzekeringsraad, Zoetermeer, 1993.
KLOSSE, S., *Menselijke schade: vergoeden of herstellen? De werking van (re)integratieregelingen voor gehandicapten in de Bondsrepubliek Duitsland en Nederland*, Antwerpen/Apeldoorn: MAKLU, 1989.
KOTZ, H. & WAGNER, G., *Deliktsrecht*, Neuwied: Luchterhand, 2001.
LERMAN, K.A., Bismarckian Germany, Chapter 6, in: Breuilly, J. (ed.), *Nineteenth Century Germany. Politics, culture and society 1780–1918*, London: Arnold, 2001.
LINDENBERGH, S.D., *Arbeidsongevallen en beroepsziekten*, Deventer: Kluwer 2000.
MAGNUS, U., Compensation for Personal Injuries in a Comparative Perspective, 2000 *Wasburn Law Journal* 39.
MARKESINIS, B.S., DEAKIN, S. & JOHNSTON, A., *Markesinis and Deakin's Tort Law*, Oxford: Clarendon Press, 2003.
MARKESINIS, B.S. & UNBERATH, H., *The German Law of Torts. A Comparative Treatise*, Oxford: Hart Publishing, 2002.
NATIONAL BUREAU OF ECONOMIC RESEARCH, INC., *Insurance Rationing and the Origins of Workers' Compensation*, Working Paper No. 4943, Cambridge: National Bureau of Economic Research, 1994.

NIPPERDEY, TH., _Deutsche Geschichte 1866–1918, Band I. Arbeitswelt und Bürgergeist_, München: Verlag C.H. Beck, 1998.

PELSTER, H.J.W., _Ongevallenverzekering. Eene Inleiding tot de Studie van Wetten betreffende de Schadevergoeding aan Werklieden ter zake van Bedrijfsongevallen, in het bijzonder tot de Studie van de Nederlandsche Wet_, Leiden: Futura, 1906.

RAUWS, W., Financiering van schade veroorzaakt door arbeidsongevallen en (nieuwe) beroepsziekten: België als wenkend voorbeeld?, in: _Schade door arbeidsongevallen en nieuwe beroepsziekten_, Faure, M. & Hartlief, T. (eds.), Den Haag: Boom Juridische Uitgevers, 2001, p. 109–129.

VAN REYBROUCK, P., _De gezondheidszorg in de arbeidsongevallen- en beroepsziektenverzekering in de lid-staten van de Europese Unie_, Antwerpen/Apeldoorn: MAKLU, 1995.

SCHOKKAERT, E., Altruism, efficiency and justice: ethical challenges to the welfare state, in: Schokkaert, E. (ed.), _Ethics and Social Security Reform_, Aldershot a.o.: Ashgate, 2001, pp. 3–35.

SCHWITTERS, R.J.S., _De risico's van de arbeid. Het ontstaan van de Ongevallenwet 1901 in sociologisch perspectief_, Groningen: Wolters-Noordhoff, 1991.

SEELEIB-KAISER, M., The Welfare State: Incremental Transformation, Chapter 7, in: Padgett, S., Paterson, W.E. & Smith, G. (eds.), _Developments in German Politics 3_, Basingstoke: Palgrave Macmillan, 2003.

TINGA, K. & VERBRAAK, E., Solidarity: An indispensable concept in social security, in: Van Vught, J.P.A. & Peet, J.M. (eds.), _Social Security and Solidarity in the European Union. Facts, Evaluations, and Perspectives_, Heidelberg/New York: Physica-Verlag, 2000, pp. 254–269.

Shifts in Work-Related Injuries: An Explanatory Analysis

R.I.R. Hoop*

1. Introduction

There are different ways in which damages caused by work-related injuries or occupational diseases can be compensated. The main compensation systems in this field consist of social security and civil liability law. The first is public, the latter is private in character. Obviously, all kinds of intermediate forms, variations and combinations of these two systems are conceivable and can be encountered in practice. The contribution of Engelhard in this book shows in addition that in all the examined countries the governmental policy regarding the compensation of work-related injuries and occupational diseases shifts from one system to another in due course.[1] Compensation systems seem to be interchangeable then. However, this does not mean that public and private compensation systems operate as communicating vessels. For each system disposes of it's own characteristics and basic assumptions by which it distinguishes itself from another.[2] One can only make mention of a complete interchangeability in so far as the essential characteristics, the basic assumptions typical of a system are sufficiently maintained in the alternative compensation system.[3]

1

---

* Researcher at the University of Maastricht and at the Social Law Department of the Vrije Universiteit Brussel (VUB). This contribution also makes part of the research project 'Interdisciplinaire studie van de verplichting tot het verschaffen van arbeid en bestaanszekerheid. Naar een juridisch denkkader voor de verdeling van verantwoordelijkheid tussen samenleving, opdrachtgever en kostwinner', financed by the Research Foundation – Flanders (nr. G.0028.04).
[1] This does not imply that different compensation systems can't exist or operate next to each other at the same time. At the contrary, as Philipsen rightly points to in his contribution, compensation systems often consist of 'different layers' of compensation. Each layer may involve a different compensation system.
[2] S. Klosse, Schadevergoeding via sociale zekerheid en aansprakelijkheidsrecht: communicerende vaten?, in: M. Faure/T. Hartlief (eds.), *Schade door arbeidsongevallen en nieuwe beroepsziekten* (2001), 13; M.G. Faure/T. Hartlief, *Nieuwe risico's en vragen van aansprakelijkheid en verzekering* (2002), 212.
[3] S. Klosse (*supra* fn. 2), 4; S. Klosse/G. Vonk, De betekenis van het recht voor de toekomst van de sociale zekerheid, in: S. Klosse (ed.), *Sociale zekerheid: een ander gezichtspunt. Toekomstperspectief vanuit vier disciplines* (2000), 210.

2   This finding is important for the research that I intend to carry out, for it raises
     two questions with regard to the evaluation of the observed shifts[4] in gover-
     nance. One, was there an awareness of the differences between compensation
     systems when shifting from one to another? And two, if so, which of these dif-
     ferences have then affected an (controlling) influence on the choice for a cer-
     tain compensation system? Indeed, one may assume that being aware of the
     fact that compensation systems have distinct characteristics and principles,
     governments will not have made shifts arbitrarily. Before analyzing the con-
     siderations and motivations for the observed shifts from this perspective, it
     seems useful therefore to give a short overview of the main purposes and pe-
     culiarities of the different compensation mechanisms.

2.   Characteristics and Functions of Compensation Systems

2.1   Compensation

3   As mentioned in the introduction two main compensation systems can be dis-
     tinguished: tort (liability) law and social security. Obviously, the two systems
     have at least one purpose in common, compensating for damages. Moreover,
     they both represent a departure from the rule that the losses lie where they fall
     (*casum sentit dominus*) and offer a justification for a displacement of the bur-
     den of damage.[5] The actual content of this justification differs however. In the
     case of tort law it is considered unfair or unjust that someone should suffer a
     loss that was caused by somebody else.[6] Corrective justice demands that the
     balance of fairness that the defendant has upset by negligence or by creating a
     risk of injury is redressed.[7] The compensation here is corrective and essential-
     ly backward-looking: It aims to restore the plaintiff in the position he was in
     before the tort occurred.[8] On the one hand this reasoning implies full compen-
     sation. On the other hand it also entails important limitations. Indeed, whether
     it concerns strict or fault liability some sort of shortcoming or failing on behalf
     of the defendant will have to be proven in order to make him pay for the loss-
     es.[9]

4   This kind of restrictions or limitations are absent in the case of social security
     law. The barriers to enter the system are much lower. The foundation for the
     displacement of the burden of damage from the individual to the collective
     level here lies in the felt necessity to guarantee a certain standard of protection

---

[4] With regard to the criteria as to when a change in an existing compensation mechanism will be
     regarded as a shift, see the introduction by Klosse and the contribution of Engelhard.
[5] Logically, a third compensation system consists of first party insurance, this is the situation
     where the losses stay where they fall and where the victim protects himself via private insur-
     ance.
[6] S. Klosse/G. Vonk (*supra* fn. 3), 200.
[7] P. Cane, *Atiyah's Accidents, Compensation and the Law* (6th edn. 1999), 359.
[8] P. Cane (*supra* fn. 7), 350.
[9] P. Cane (*supra* fn. 7), 359; T. Hartlief, *Ieder draagt zijn eigen schade. Enige opmerkingen over
     de fundamenten van en ontwikkelingen in het aansprakelijkheidsrecht* (1997), 26; S. Klosse/G.
     Vonk (*supra* fn. 3), 200–201.

by the government, based on the conviction that certain damages represent a social rather than an individual risk.[10] The idea is that these kinds of risks and costs should be spread as wide as possible, so that the protection comes to include the so-called 'bad risks' as well. This explains why premiums are mostly income-related and insurance is often compulsory, for these elements of solidarity are necessary in order to balance the costs between 'good' and 'bad' risks.[11] There is however a price to be paid for this solidarity. Social security rarely leads to full compensation (as is the case in the liability system), but offers a fixed and lower level of protection set by the government.[12] The compensation here is more redistributive in character and essentially forward-looking. 'It is not concerned with making up for the past but with improving people's lives in the future. The social security system, (...), is not concerned with how a person came to be in the position they are in but whether fairness and humanity demand amelioration of that position'.[13]

Both systems clearly have a compensation function then. Still, the presented justification is different and the compensation differs in character: Tort liability law offers full compensation once strict conditions are met; social security offers an easier access but at a lower compensation level.[14]   5

2.2   Prevention, Loss Distribution and Cost-Allocation

Apart from compensating for damages, compensation systems are generally also associated with some other functions of which prevention, loss distribution and risk or cost-allocation are the most important ones for our subject.[15]   6

Prevention is evidently of great importance in relation to industrial accidents. While originally social policy was confined to or mainly focused at compensating for damages, more and more attempts are made to equally concentrate on repairing and even preventing damages.[16] Prevention or deterrence especially comes to the fore in connection with liability rules. The basic idea is that the prospect of having to pay damages for injuries caused by particular conduct will deter people from engaging in conduct of that type.[17] Liability rules are thus seen as rules of conduct. It is the economic analysis of law that   7

---

[10] S. Klosse/G. Vonk (*supra* fn. 3), 196–197; G.A. Ritter, *Social Welfare in Germany and Britain: origins and development* (1983), 5.
[11] S. Klosse (*supra* fn. 2), 4–5.
[12] S. Klosse/G. Vonk (*supra* fn. 3), 198–200; T. Hartlief (*supra* fn. 9), 29; M.G. Faure/T. Hartlief (*supra* fn. 2), 210–211.
[13] P. Cane (*supra* fn. 7), 350.
[14] M.G. Faure/T. Hartlief (*supra* fn. 2), 211–213. The same principle seems applicable with regard to strict liability where the easier recovery for the victims is often accompanied by less compensation, see U. Magnus, Compensation for Personal Injuries in a Comparative Perspective, [2000] *Washburn Law Journal* (WLJ) 39, 355.
[15] P. Cane (*supra* fn. 7); T. Hartlief (*supra* fn. 9), 16–17.
[16] See S. Klosse, *Menselijke schade: vergoeden of herstellen? De werking van (re)integratieregelingen voor gehandicapten in de Bondsrepubliek Duitsland en Nederland* (1989).
[17] P. Cane (*supra* fn. 7), 361.

has taken this idea as its starting point.[18] Legal rules are approached as encouragements, incentives for careful conduct, which can increase social safety.[19]

8   It is apparent that the tort system by linking liability to pay compensation with responsibility for causing accidents, may, to some extent, further the goals of general deterrence. Still, as Cane rightly points out, the general deterrence potential of tort law is also limited.[20] First of all, the prevalence of liability and first party insurance greatly reduces the deterrent potentiality of tort law since insurances are a possibility to spread the risk and distribute losses.[21] Here a conflict arises with the compensation goal, which is on the contrary just served by the availability of insurances. A second problem is the fact that the concept of 'causality' as it exists in civil liability not always or automatically coincides with the basic assumption of the theory of general deterrence that holds that accident costs should be borne by the person who can most cheaply avoid accidents of that type in the future.[22] Thereupon the tort system does not impose all the costs of accidents and does only take account of private losses.[23] The administrative costs of tort law are also relatively high.[24]

9   The possible preventive effects of social security arrangements seem to be even more limited. The central role of solidarity and the still dominant focus on redistributive compensation that have shaped these arrangements are difficult to reconcile with the demands urged by notions such as causality and prevention.

10   The spreading of losses makes them more tolerable for all. The idea of loss distribution offers an attractive new justification for tort law. The effect of the tort system is indeed not, in general, merely to shift a loss from one person to another. There is an important side effect: The combined effect of tort law, liability insurance and the operation of the market is, in practice, to distribute losses among a large number of people and over a period of time.[25] However there are much cheaper and more efficient ways of distributing the losses; with regard to administrative costs for example tort law appears more expensive than social security arrangements.[26] In the case of work-related injuries losses can be spread among a collectivity of insured[27] or by passing them on to employees (lower wages), shareholders (lower dividends), consumers (higher

---

[18] For a more elaborate overview of this perspective, see the contribution of Philipsen.

[19] T. Hartlief (*supra* fn. 9), 19.

[20] P. Cane (*supra* fn. 7), 383–384.

[21] P. Cane (*supra* fn. 7), 368.

[22] P. Cane (*supra* fn. 7), 376.

[23] P. Cane (*supra* fn. 7), 385.

[24] P. Cane (*supra* fn. 7), 338.

[25] P. Cane (*supra* fn. 7), 354–355. As Magnus remarks tort law now functions to a large extent as a means to allocate costs between insurance companies: U. Magnus [2000] WLJ 39, 358.

[26] P. Cane (*supra* fn. 7), 355–356.

[27] T. Hartlief (*supra* fn. 9), 19. A further choice can be made between private or state-run first party loss insurance (spreads among potential victims) or third party insurance (spreads amongst those likely to inflict it); P. Cane (*supra* fn. 7), 355.

prices) and even tax-payers (as deductible expenses).[28] The spreading of loss-es is clearly an objective within social security regimes where solidarity is a key principle.[29] The range of distribution can vary from a specific group of workers to all citizens of a state.

A final possible function is the allocation of risks or costs. This function can   11 be connected with two earlier functions. Costs or risks can be allocated to the party that is most equipped to compensate for the damages (cf. compensation), to the party that is best suited to spread the loss[30] (cf. loss distribution) or the allocation of costs can have an indirect preventive effect in that certain activi-ties will be less exercised due to their increased cost (cf. prevention).[31] This latter view seems to have inspired systems of strict liability under tort law. In cases of fault liability 'responsibility' appears to be the criterion of allocation. In most social security arrangements there is no clear criterion of allocation. The general idea is that costs and risks should be spread as much as possible in order to establish an equal protection for the so-called 'bad' risks (cf. soli-darity). This kind of allocation can be realized by rendering insurances man-datory or by making premiums income-related.[32] The risk-relatedness of pre-miums is then a departure from this principle and refers to another criterion of allocation.

3. Considerations and Justifications for the Shifts

Given the differences between compensation systems, shifts from one system   12 to another will as a rule not being made arbitrarily. On the contrary, in particu-lar with regard to the significant policy changes, i.e. the real shifts that consti-tute the actual subject of the shifts-project, one may assume that the various peculiarities of compensation systems precisely constitute the main cause to (even) consider a change. This also means that the reasons given to justify a shift will probably not only refer to the expectations one has vis-à-vis the fu-ture compensation system, but may also reveal the deficiencies of the former system and the elements on the basis of which it was judged unsatisfying from a certain moment on.

The description of the reasons and arguments that (may) have inspired the var-   13 ious shifts in government with respect to the compensation of work-related in-juries is the topic of this third chapter. I will systematically go into the under-lying motives for the various policy shifts as they were developed and described by Engelhard, thereby inevitably repeating some of the elements that she already touched upon. The arguments explicitly put forward during the parliamentary proceedings of the legislative changes are thereby taken as a

---

[28] P. Cane (*supra* fn. 7), 339.
[29] S. Klosse (*supra* fn. 2), 4–5; S. Klosse/G. Vonk (*supra* fn. 3), 199. Cf. *supra*.
[30] P. Cane (*supra* fn. 7), 357.
[31] T. Hartlief (*supra* fn. 9), 21.
[32] S. Klosse (*supra* fn. 2), 4–5; S. Klosse/G. Vonk (*supra* fn. 3), 199.

Functions	Liability Law (Tort)	Social Security
Compensation	Yes, corrective (losses) • full compensation possible • high thresholds (causality)	Yes, redistributive (needs) • limited in time and scope • low thresholds (irrespective of cause; solidarity)
Prevention	Liability rules as incentives for careful behaviour (cf. Law & Economics): a) *fault-*: standard of care is given; evaluation by a judge *ex post* b) *strict-*: standard of care to be evaluated by the employer *ex ante* BUT: limitations! 1. because of insurance possibility (conflict compensation – prevention) 2. causality ≈ possibility to avoid costs the cheapest way possible 3. high administrative costs 4. no social losses in calculation 5. not all accident costs taken into account	Very limited: 1) primary goal = compensation 2) central role for principle of solidarity Possibilities with regard to work-related injuries: via premium differentiation, rehabilitation efforts
Spreading of risks	Yes, especially due to insurance possibility (collectivity of insured) + transfer through the market (consumers/employers/share-holders) + on government (via deductible expenses).	Yes, because of solidarity mechanisms (premiums and benefits) and principle of spreading! Going from a limited group of employers to all residents.
Cost allocation	Yes, strict liability	Not clear: maximum spreading (departure when premium differentiation)

point of departure. Other sources are the recommendations and reports made by advisory committees and of course legal and historical literature.

3.1 The Shifts in Germany

3.1.1 The First Shift

As in most West European countries at the end of the 19th century, the rise of capitalism and the continuing process of industrialisation together with accompanying tendencies such as urbanisation, the professionalisation of labour relations and the desintegration of traditional family ties, gave rise to the so-called 'social question', i.e. the problem of the integration of the industrial workers into the existing social and political order. In the industrialized economy workers had to protect themselves against economic hardship in case of sickness or old age, but generally lacked the means thereto. The industrial development had moreover created a new and important category of economic hardship, namely the disability to work caused by a work-related injury.[33]   14

All these developments and the social unrest they brought into existence made the necessity clear of a more adequate worker's protection. According to the existing civil law rules the employee-victim of an industrial accident could only obtain compensation for his injuries if he was able to proof that his damage was the result of the fault or negligence of the employer. If the damage was caused by a representative, then the employer was only responsible for the *culpa in eligendo*.[34] In practice this meant that only very few victims succeeded in delivering this evidence,[35] and when they did, there was always the risk of being confronted with the insolvency of the employer.[36] Especially in the dangerous mining industry the demand for a better compensation scheme was strong.   15

In April 1868 a petition from Leipzig proceeding from Liberal quarters, in which a revision of the legal compensation regime with regard to accidents in factories, mines and railways was asked, was presented to the Reichstag (German parliament).[37] This initiative resulted in the promulgation of the *Reichs-*   16

---

[33] W. Gitter, *Schadenausgleich im Arbeitsunfallrecht. Die soziale Unfallversicherung als Teil des allgemeinen Schadensrechts* (1969), 5–6; G.A. Ritter (*supra* fn. 10), 1–3.

[34] See Explanatory memorandum, in E. Wickenhagen, *Geschichte der gewerblichen Unfallversicherung* (1980), 7; W. Gitter (*supra* fn. 33), 11. At least, this was the case in the majority part of the German empire, but not in the Rhineland that used the French Civil Code. See Engelhard; J.M. Kleeberg, From Strict Liability to Workers' Compensation: the Prussian Railroad Law, the German Liability Act, and the Introduction of Bismarck's Accident Insurance in Germany, 1838–1884, [2003] *Journal of International Law and Politics* (JILP) 36, 63–65 and 88–89.

[35] The formal rules of proof to which the German legal system adhered at that moment constituted another obstacle: W. Gitter (*supra* fn. 33), 12–13.

[36] S. Klosse (*supra* fn. 16), 13.

[37] W. Gitter (*supra* fn. 33), 14; S. Klosse (*supra* fn. 16), 13.

haftpflichtgesetz[38] *of 7 June 1871* (the German Liability Act of 1871,
RHpflG). This Act made railway operators liable for all the accidents occurring during the exploitation, unless the employer could proof that the accident
was caused by an Act of God or the victim's own fault. This clause that installed a presumption of fault on account of the employer, was in fact a copy
of the *Prussian Railroad Act* that already dated from 3 November 1838.[39] For
accidents occurred in one of the other '*mit ungewöhnlicher Gefahr verbun-
denen Unternehmungen*' such as mines and quarries, the employer would only
be responsible if the victim or his bereaved could proof the fault of the employer
or – and this vicarious liability was the new element – one of his agents.

17   It was the retention of this heavy burden of proof as well as the still limited
     scope of application of the Act[40] that explains why the RHpflG still did not
     give any consolation to victims of work-related injuries.[41] The Act provoked
     moreover more and protracted lawsuits, especially when insurance companies
     got involved. These latter waited until the matter had been taken to court and
     then offered the victim-employee a poor settlement. These practices obviously
     did not improve the labour relation and both employers and workmen were
     dissatisfied with the results of the Act.[42]

18   The discussion concerning the social question took on new life from the late
     1870s onwards due to the serious economic depression after 1873. The confi-
     dence in the self-regulating capacity of the free market was shocked and the
     fear for a potentially revolutionary proletariat increased.[43] Since the introduc-
     tion of the direct and equal parliamentary male suffrage in 1866–1867 and the
     proclamation of the *Koalitionsrecht* (the right of association) in 1869, the way
     had been paved for the rise of the labour[44] and trades union movement. This
     resulted in frequent strikes during the period from 1869 to 1874.[45] In view of

---

[38] *Reichshaftpflichtgesetz betreffend die Verbindigkeit zum Schadenersatz für die bei dem
      Betriebe von Eisenbahnen, Bergwerken, Fabriken, Steinbrüchen und Gräbereien herbeige-
      führten Tödtungen und Körperverletzungen of 7 June 1871* (Law Concerning the Obligation to
      Compensate for Damages Resulting from Deaths and Injuries Caused by the Operation of Rail-
      roads, Mines, etc.).
[39] G. De l'Hôpital, *L'assurance contre les accidents du travail en Allemagne* (1904), 9; W. Gitter
      (*supra* fn. 33), 15; J.M. Kleeberg, [2003] JILP 36, 66–81; J. Van Steenberge, *Schade aan de
      mens. Deel I. Evaluatie van de arbeidsongeschiktheid in het recht* (1975), 87.
[40] That a more comprehensive reform (*e.g.* the adoption of the principle of vicarious liability for
      the whole of liability law) was dissuaded, was due to a form of legal conservatism that wished
      to hold on to the traditional 'fault-principle': W. Gitter (*supra* fn. 33), 14 *et seq.*
[41] J. Van Steenberge (*supra* fn. 39), 88.
[42] G. De l'Hôpital (*supra* fn. 39), 9–10; J. Van Steenberge (*supra* fn. 39), 121; E. Wickenhagen
      (*supra* fn. 34), 29.
[43] G.A. Ritter (*supra* fn. 10), 1.
[44] In 1869 the *Sozialdemokratische Partei Deutschlands* (Social Democratic Party, SPD) was
      founded. In the 1874 and 1877 elections for the Reichstag the socialists succeeded in winning
      about 9% of the votes, and in the strongly industrialised and urbanised Protestant areas, they
      even became the strongest party with more than one third of the votes. G.A. Ritter (*supra*
      fn. 10), 24 and 31–32.
[45] G.A. Ritter (*supra* fn. 10), 25.

this manifest threat to the existing social and economic order, the forces in society who wanted to avoid an open conflict, came in action in pursuit of social reforms. The Christian-Social movement (with the Centre Party as spokesman) had the purpose of preventing the workers drifting away from Christianity and the Church and stressed the importance of the Christian family, the local community, corporatism and the subordinate role of the state.[46] Also the liberal professions and the higher civil service[47] were in favour of the idea that the lower classes had to be educated to self-help via their support to the existing saving associations and benefit societies, but they also believed that the State had to take up its task as the guardian of property and public order through the implementation of social reforms.[48] On the academic level attention for social reforms was asked by the so-called *Kathedersozialisten.*[49]

Also Chancellor Bismarck regarded concrete social reforms as a means to counter the success of the SPD and to reconcile the labour force with the German State. Before he had tried to eliminate the labour and trades union movement in a repressive way by an Act of 21 October 1978, which he had adopted after the attempted murders of the Emperor.[50] But this policy had an adverse effect and soon he realised that in order to act effectively against the excesses of socialism, a positive reaction was necessary directed at the welfare of the labourers.[51] Bismarck hoped that a policy of social reforms would have a state-stabilizing effect.[52] In fact he tried, via a social policy from above that would guarantee the labourers an income when they were excluded from employment, to make up for the lack of legitimacy in the political domain and integrate the workmen in bourgeois society without awarding them equal rights of political participation.[53]   19

The first proposal regarding work-related injuries of 8 March 1881 consisted in the retention of the statutory regulations in place, but with the obligation for the employer to insure. The *Reichstag* could agree with the insurance obligation but opposed to the state's role in financing and administering the sys-   20

---

[46] G.A. Ritter (*supra* fn. 10), 25–26.
[47] Due to Germany's relatively late industrialisation, its political liberalism was greatly influenced by these groups who, contrary to the commercial classes and industrial bourgeoisie, did not perceive unrestrained economic growth and industrialisation as unequivocal positive developments, but rather feared a potentially revolutionary proletariat. G.A. Ritter (*supra* fn. 10), 18.
[48] G.A. Ritter (*supra* fn. 10), 18.
[49] G.A. Ritter (*supra* fn. 10), 27.
[50] G.A. Ritter (*supra* fn. 10), 28–29.
[51] This idea was first expressed in a Royal Speech of 15 February 1881 at the opening of the new session of the *Reichstag*, and would be repeated in Bismarck's Royal Message of 17 November 1881.
[52] E. Wickenhagen (*supra* fn. 34), 36.
[53] According to some historians this perspective would explain why the authoritarian governments took the lead over the parliamentary democracies in establishing social insurance regimes since they would have had a greater need to take the initiative in this field in order to defend the system against the workers' political mobilisation. Cf. G.A. Ritter (*supra* fn. 10), 3 who refers to Alber.

tem.[54] They strongly objected to what they interpreted as *Staatssozialismus* (State socialism). Moreover one questioned if the extension of the employers' civil liability to a strict liability for all accidents would not be to great a burden on the industry and detrimental to their international competitiveness.[55] Because Bismarck did not want to give up the government contribution (which he considered essential to obtain the abovementioned political goal), he ultimately let the *Bundesrat* (the Federal German Upper House) reject the proposal.[56]

21 In his celebrated *Kaiserliche Botschaft* (Royal Message) of 17 November 1881,[57] Bismarck repeated the necessity of positively complementing the anti-Socialist laws and envisaged three mandatory social insurances: accident insurance, health insurance, and old age and disability insurance.[58] Bismarck left no doubt as to the public character of the scheme. In comparison with the legal proposal of six months earlier, a shift could be noticed from private to public law: The principle of the employer's liability was replaced by an insurance obligation pertaining to public law.[59] Bismarck also made clear that the insurances would have to be carried out by publicly organised corporations.

22 The introduction of social insurances was not only a means of thwarting the socialist success, but also a way to partially replace poor relief, which was – by its mandatory character – a heavy financial burden to the community and perceived as humiliating by its beneficiaries. Long before economists had already suggested insurances as a solution for this.[60] Insurance which had started in the 19th century as a rather new phenomenon had moreover become ever more popular.[61] The RHpflG had evidently strengthened that trend. The idea of a mandatory 'social' insurance met relatively little resistance. This is generally explained for by the weak liberalism in Germany, the already existing authoritarian state structure with a powerful and confident bureaucracy, and the long tradition with regard to the idea that the state had a special role in promoting social welfare.[62]

23 On 8 March 1882 a second proposal with regard to the compensation of work-related injuries was presented to parliament. For the first 13 weeks of disability compensation would be provided by a mandatory health insurance organ-

---

[54] G. De l'Hôpital (*supra* fn. 39), 13; S. Klosse (*supra* fn. 16), 12. Insurance was to be taken via the *Reichsversicherungsanstalt* (Imperial Insurance Institute) and private insurers were not allowed as Bismarck wanted to avoid private speculation (S. Klosse (*supra* fn. 16), 15; E. Wickenhagen (*supra* fn. 34), 33). The costs of this scheme were to be born by the employers (two thirds) and via a government subsidy (one third). No contribution was asked from the workmen themselves.

[55] E. Wickenhagen (*supra* fn. 34), 36.

[56] E. Wickenhagen (*supra* fn. 34), 35–38.

[57] E. Wickenhagen (*supra* fn. 34), 40–41.

[58] As legitimacy Bismarck referred among other things to the Christian moral basis of the state.

[59] J. Van Steenberge (*supra* fn. 39), 102.

[60] G. De l'Hôpital (*supra* fn. 39), 10–12.

[61] J. Van Steenberge (*supra* fn. 39).

[62] G.A. Ritter (*supra* fn. 10), 17–19.

ised via the *Krankenkassen* (National Health Service) to which both employers and employees would have to pay contributions. After this period or in case of a fatal accident compensation would be paid on the basis of a mandatory work-related injuries insurance (to which workers paid no contribution) by the so-called *Berufsgenossenschaften* (industrial insurance boards, BGs), mandatory, state supervised associations formed on the basis of industrial branches in which both sides of industry would be equally represented. The administration of the scheme through these corporate associations implied a certain degree of autonomy for employers and employees and was seen by these as a necessary counterbalance against the mandatory character of the insurance.[63] The employers' premiums would be linked to the risk they represented with regard to the occurrence of industrial accidents in their branch of industry.

After Bismarck had repeated in his Royal Message of 14 April 1883 the urgent character of establishing an industrial injuries insurance (referring again to industrial peace) and after the *Krankenversicherung* (Health Insurance Act) of 15 June 1883 had passed,[64] a third and final proposal was put before parliament on 6 March 1884. The range of application was reduced to that of the RHpflG[65] and superior to the BGs came now a *Reichsversicherungsamt* (Imperial Insurance Office, RVA) that would be the final instance in settling disputes. Employers were to bear the full cost of the scheme; there was no overt[66] state subsidy anymore. This proposal would lead to the *Unfallversicherungsgesetz of 6 July 1884* (Industrial Injuries Insurance Act, UV) that would come into force on 1 October 1885.   24

The UV of 1884 was based on the idea of *risque professionnel* (occupational risk). Damages caused by industrial accidents were considered to be part of the business risk and should therefore be borne by the employer regardless of any fault. Since compensation was to be disconnected from the question of guilt and employers were not expected to pay twice, civil liability law was largely put aside. Except for cases of intent and gross negligence, the employer was granted immunity against civil actions.[67] Civil immunity also had to serve social peace.[68] Also other features of the final scheme reflected an element of compromise: benefits were limited up to 66% of the last wage and the compensation for medical treatment was fixed. Under pressure of the insur-   25

---

[63] S. Klosse (*supra* fn. 16), 22.
[64] The enactment of this Act would have been less controversial because rudimentary health insurance programs already existed on the municipal level. J.M. Kleeberg, [2003] JILP 36, 111.
[65] However, in the Explanatory memorandum an extension was envisaged that already would take place in 1885. See E. Wickenhagen (*supra* fn. 34), 51–52.
[66] There was however an indirect financial contribution of the State as it was the government that paid for the operating costs of the RVA and the Post Office that advanced the payments.
[67] S. Klosse (*supra* fn. 16), 15–16.
[68] B. Barentsen, *Arbeidsongeschiktheid. Aansprakelijkheid, bescherming en compensatie* (2003), 139.

ance companies no compensation would be provided for other material damages nor for moral damage.[69]

26 Although Germany was the first country to establish this kind of mandatory state organised insurance, the system was well received by the employers who adopted a fairly cooperative attitude towards it.[70] The German compensation scheme and its early results would soon receive attention and admiration abroad.[71] Consequentially, Germany took great pride in its exemplary role to other nations.[72]

3.1.2 Later Developments

27 The basic structure of the German compensation scheme for work-related injuries has largely remained unchanged to this day. The system, in which the BGs have played a central role in taking the initiative for further improvements and developments, has indeed proven to be able to bear the test of time and to absorb technical evolutions and times of depression. The scheme is deemed to have attained its full development, to operate inconspicuously and to serve social peace. It is moreover believed to be able to remain playing its role of pioneer in the future.[73]

28 In the first decades[74] following the UV of 1884 the field of application of the scheme was further extended. In 1925 the notion 'industrial accident' came to comprise certain occupational diseases as well as commuting accidents. The insurance coverage was also gradually expanded from the so-called 'dangerous' to the non-dangerous industries, as well as to commercial and administrative activities. The already by Bismarck envisaged[75] full range of insured was reached in 1942, when the UV evolved from a business insurance to a personal insurance: not the type of business but the labour relation became determining as to the applicability of the scheme. The scheme became further detached from industrial or labour activities when under the same insurance coverage was provided for lifesavers, students, volunteers, donation of blood and tissue and similar socially welcome activities.[76] This part came to be known as the *unechte Unfallversicherung* (the unreal accident insurance).

29 At an early stage attention was paid by the BGs to the field of accident prevention and rehabilitation, a practice which was only afterwards provided with a legal basis. The activities in these areas were especially expanded after the second World War when in all West European countries the ideal of a welfare

---

[69] S. Klosse (*supra* fn. 16), 22; J. Van Steenberge (*supra* fn. 39), 121.

[70] G. De l'Hôpital (*supra* fn. 39), 38 and 162.

[71] *E.g.* G. De l'Hôpital (*supra* fn. 39), 162–163.

[72] G. De l'Hôpital (*supra* fn. 39), 61.

[73] E. Wickenhagen (*supra* fn. 34), 442–443.

[74] In 1911 a uniformization of the existing social insurances had taken place by which the UV had become book 3 of the *Reichsversicherungsordnung* (State Insurance Code, RVO).

[75] E. Wickenhagen (*supra* fn. 34), 279.

[76] See § 2 (1) *Sozialgesetzbuch* VII (Social Code, SGB).

state was enthusiastically pursued. This resulted in expenditures for hospitals, first aid posts, research institutes et cetera. Prevention and rehabilitation were embedded as main objectives with the introduction of the *Unfallversicherungs-Neuregelungsgesetz* (New Industrial Injuries Insurance Act, UVNG) in 1963.[77] The BGs gained the authority to sanction the *Unfallverhütungsvorschriften* (accident prevention instructions), regulations issued with a view to accident prevention. Employees could incur a fine; employers saw their premium increased or lowered in function of the accident history of their individual business.[78] Prevention was further stimulated through other Acts such as the *Maschinenschutzgesetz of 1968* (Machine Protection Act) and the *Arbeits-sicherheitsgesetz of 1973* (Labour Security Act).

In the 1970s and 80s the expansion of social policy came to an end and cost   30
containment came to the fore due to the economic depression, although the work-related injuries section relatively got away unscathed, since the number of accidents was decreasing.[79] Still, some BGs struggled with financial problems.[80] The work-related injuries scheme became formally more closely integrated in the global social security system when it became part of the in 1975 introduced *Sozialgesetzbuch* (Social Code, SGB) that tried to bring unity and harmony in the existing social insurances without changing them.

The aspiration to further unify and uniformize the different social insurances   31
branches (as a consequence of a general evolution in minds from causality to finality) has remained a threat to the unique position of the Industrial Injuries Insurance Act, although so far the scheme has succeeded in safeguarding its current position. Another matter which has often been discussed, is the question whether accident prevention and rehabilitation should be organised in separation of the compensation scheme. A modification in that sense would mean abandoning one of the basic principles of the scheme and probably one of the most successful principles since it is believed that the conjunction of prevention, rehabilitation and compensation together with the central role of the BGs with regard to initiative and responsibility, can be seen as the determining factors for the success of the system.[81]

3.2   The Shifts in the Netherlands

3.2.1   The First Shift

The introduction of the *Ongevallenwet* in 1901 (Industrial Injuries Act 1901,   32
OW) constitutes for the Netherlands a first important shift in the governmental policy with regard to the compensation of damages caused by work-related injuries.

---

[77] § 537 RVO UVNG 1963 (§ 1 SGB VII).
[78] In practice not much use is made of these possibilities: B. Barentsen (*supra* fn. 68), 127.
[79] S. Klosse (*supra* fn. 16), 454 and 462.
[80] E. Wickenhagen (*supra* fn. 34), 432.
[81] E. Wickenhagen (*supra* fn. 34), 441.

33   The primary goal of this legislative initiative was very distinct and explicit: of-
      fering more protection to the victims of this type of injuries. The injured
      worker had to be made sure of compensation in case of an industrial acci-
      dent.[82] In the explanatory memorandum of the Act, a very reference is made to
      'a natural right on compensation', a right that the government should guaran-
      tee. Also most lawyers at the time shared the conviction that the possibility for
      the victim to get indemnified should no longer be dependent of the answer to
      the question who was to blame for the accident.[83]

34   The crux of the criticism on the civil liability system, the compensation sys-
      tem injured workers were committed to before the OW, indeed touched upon
      the fault principle.[84] As mentioned above, demanding for compensation from a
      party that is hold responsible implies that a fault or some sort of shortcoming
      can be proven on behalf of this party. In the late nineteenth economy with its
      complex and impersonal industrial production methods, this was considered
      an increasingly difficult exercise. The modern industry gave rise to accidents
      which escaped the fault – Act of God dichotomy[85] and also made it more diffi-
      cult to invoke the vicarious liability of the employer, since this latter did not
      choose nor supervise his own employees any longer. At the same time, the
      standard defenses of contributory negligence[86] and of *volenti non fit injuria*[87]
      to which the employer could make an appeal in this system, were no longer
      considered justified. The workman was in no position to value the risks and
      dangers of the industrial workplace, and some carelessness (caused by habitu-
      ation) appeared to a certain extent necessary and seemed beneficial to the en-
      terprise. In a word, the burden of proof on the accident victim weighed too
      heavy.

35   In addition to these legal obstacles came some financial and social barriers.
      Starting a lawsuit was expensive and could later on prove to be counterproduc-
      tive taking into account that the employee's subsistence depended upon wage
      labour. Legal procedures could also threaten the social peace in the enterprise,
      especially when other employees were involved as witnesses. And of course
      there was always the risk of insolvency on the part of the employer. In general,
      it was unimaginable that workers would take legal action against their em-
      ployers. Civil liability law was only considered applicable for the equally
      prosperous. The possibility of holding an employer legally responsible existed

---

[82] Kamerstukken I, 1899–1900, 221.
[83] R.J.S. Schwitters, *De risico's van de arbeid. Het ontstaan van de Ongevallenwet 1901 in soci-
      ologisch perspectief* (1991), 252.
[84] R.J.S. Schwitters (*supra* fn. 83), 250 *et seq.*
[85] F. Ewald, *L'état Providence* (1986), 87–90; R.J.S. Schwitters (*supra* fn. 83), 290.
[86] This was the argument that no action could be allowed if it were shown that the accident victim
      was in any degree responsible for his fate. In other words, if the evidence proves that the injury
      was a consequence of the joint negligence of both parties, the injured person cannot recover
      any damages.
[87] The *volenti non fit injuria*-principle holds that in accepting dangerous employment a workman
      willingly consented to the risks involved and implicitly renounced all claims to be compen-
      sated.

only in theory, not in reality. For the workers, it was a 'non-accountable civil accountability'.[88]

The new regulation was an attempt to avoid or minimize these legal and other   36
uncertainties and obstacles in order to realize the broadly supported wish to
effectively compensate the victims of industrial accidents. The emphasis that
was put on the necessity of guaranteeing victims an adequate compensation as
the first and foremost ideal explains some important choices that were made in
form and content of the new compensation scheme, and often served as an explicit justification for these choices.

With regard to the conditions under which a right to compensation is awarded,   37
this has brought about a significant departure from the basic principles of the
civil liability system. Not the fault principle but compensation had now become the starting point. By providing a large definition of the concept 'industrial accident'[89] one tried to avoid controversies concerning the contributory
negligence of the victim and to confer this latter an easier access to damages.
Only when the worker had provoked his injuries deliberately, he was denied
the right to compensation. As a justification for this transfer from employee to
employer of the uncertainty with regard to the causality of industrial injuries,
reference was made to fairness. It was seen as unjust to let the worker bear
alone the burden of circumstances beyond one's control, coincidence or unknown causes.

Also the subordinate position of the worker was invoked as an argument. The   38
employer was considered the better positioned to take the necessary precautions. The influence the employee could exercise on his working conditions
was on the contrary minimal. Noteworthy is the fact that this last argument,
which regards the (possible) preventive function of the new regulation, only
figures in the explanatory memorandum of the first legislative proposal but
disappeared in that of the final one. Presumably the government didn't want to
stress this (new) responsibility of the employers too much in order to obtain
the agreement of the more employer-oriented members of parliament.

The OW of 1901 consisted of a compulsory collective industrial injuries insur-   39
ance borne by the employers. The choice for a regulation based on collective
insurance offered, according to the government, the best guarantee to realise
the preconceived goals and responded to the so-called principle of public law
that common dangers and expenses should also be carried in common.[90] At the
time there was a growing public support for private insurances. Non-compul-
sory private insurance created solidarity without corroding the principle of
freedom. It was therefore seen as the most appropriate instrument for the so-

---

[88] R.J.S. Schwitters (*supra* fn. 83), 6–7.
[89] The insured risk consisted of all accidents 'associated with undertaking an enterprise' (art. 1 OW; my translation).
[90] Kamerstukken II, 1898–1899, no. 16, 2. (MvA).

cial relief of workers without provoking state interference. It also accomplished the independence of the worker who would not, in case of misfortune, have to depend on poor relief or social assistance any longer.[91]

Notwithstanding the resistance amongst private insurers, a dynamic lied in the reasoning concerning the functioning of insurance, which made the possibility real for the government to set up collective and compulsory insurances. The idea for instance that private insurance would lead to independence, was easily recuperated by the government by arguing that, since private initiative proofed itself insufficient in realising this in practice, the government had to take over by organising collective and compulsory insurances.[92] The lack of private initiative at this point was indeed one of the arguments regularly put forward during the parliamentary discussions. State interference could thus be presented as serving the realisation of individual independence and responsibility.[93]

40   The compulsive character of the insurance also appeared indispensable in view of the compensation goal, to eliminate the risk of insolvency. Another reason for the choice of a compulsory collective insurance system was the belief that this system would lead to a better allocation of costs than the free market had done so far. Not 'fault' or 'guilt' as interpreted in liability law had to be the leading principle at this but the direct benefit principle: He who enjoys the profits has to bear the burdens as well.[94] Since the employer was the one who cleared the profits that were generated by the company, He was also judged the one who had to carry the risks of that production.[95] And the modern complicated production processes made belief that the risk of industrial injuries was part of that total business risk.[96] This evolution from fault to risk as the basis for allocating damages is without any doubt the most salient characteristic of the 1901 Act. It constitutes the introduction of the so-called '*risque-professionnel*'-theory in the Dutch industrial injuries regulation.[97] In that per-

---

[91] R.J.S. Schwitters (*supra* fn. 83), 210–215.

[92] R.J.S. Schwitters (*supra* fn. 83), 215–216 and 262–263; Kamerstukken I, 1899–1900, 286 and 288. (MvA).

[93] C.J.H. Jansen/C.J. Loonstra, De personele werkingssfeer van de socialever-zekeringswetten 1900–1960, in: A.Ph.C.M. Jaspers *et al.* (eds.), *De gemeenschap is aansprakelijk ... Honderd jaar sociale verzekering 1901–2001* (2001), 93.

[94] R.J.S. Schwitters (*supra* fn. 83), 260. In the same sense: Kamerstukken II, 1896–1897, 159, no. 3, 11.

[95] In spite of this, the government seemed well aware of the fact that in practice the employers had the possibility of a further spread of risk by passing on these costs to the consumer. Kamerstukken I, 1899–1900, p. 274 and 292. See also W.J.P.M. Fase, De legitimering van het verplichtend karakter van de sociale verzekering, in: A.Ph.C.M. Jaspers *et al.* (eds.), *De gemeenschap is aansprakelijk ... Honderd jaar sociale verzekering 1901–2001* (2001), 54.

[96] In the explanatory memorandum, it is said that accidents, notwithstanding all precautions, will keep on occurring and are practically inevitable and that the resulting damage should be considered a necessary cost of production, a risk intrinsic to that production. Kamerstukken II, 1897–1898, 182, no. 3, 12.

[97] P.S. Fluit/A.C.J.M. Wilthagen, Het risque social, in: A.Ph.C.M. Jaspers *et al.* (eds.), *De gemeenschap is aansprakelijk ... Honderd jaar sociale verzekering 1901–2001* (2001), 110–111; S. Klosse (*supra* fn. 16), 50–51.

spective it was considered an unjustified enrichment for the employers when the costs of industrial injuries eventually ended up being carried by the poor relief. A compulsory insurance could make an end to this practice.[98]

Cost efficiency and compensation security were also the arguments put for-   41
ward to defend the position that the regulation had to be carried out by the state and the state alone.[99] An additional argument according to the government was prevention. Private insurers could not be trusted in taking care that companies would be adequately classified in the different so called *gevarenklassen* (classes of danger) and would not be able to send out the right incentives.[100] The question whether employers had to be allowed to be their own risk-bearer or to transfer the risk to private insurers, was indeed a crucial matter during the parliamentary discussion. It was only after the abovementioned possibility was included that the *Eerste Kamer* (Upper Chamber, EK) was willing to pass the Act. The prevention argument was promptly reversed. In justifying the opportunity for the employers to bear their own risk, the government referred to the positive outcome this option would have with regard to prevention.[101]

In case of an industrial accident the employee-victim had a direct claim to   42
compensation against the *Rijksverzekeringsbank* (National Insurance Bank, RVB), a government agency, who, at his turn would recover the payments made from the employer, be it in the collection of premiums or by means of subrogation. The regulation of the 1901 Act thus transformed the private relationship between employer and employee in a double public one.[102] This public regulation guaranteed the employee a right to compensation even when the employer could not be held responsible for the accident. As a matter of balance, the OW adjudicated 'immunity' of civil liability to the employer so that the regulation also meant a gain of security for them since they now had the guarantee that possible damages were limited to the amount of the paid contributions.[103] In this way the new regulation could also be presented as a method of spreading the risks for employers: By all paying relatively low premiums employers had the security that no accident could ever have disastrous consequences for one of them.[104]

The idea that the new insurance scheme asked for compensation or restoring   43
of the balance on behalf of the employers lived quite strongly in the parliamentary discussion. The focus on establishing for the employees a right to

---

[98] R.J.S. Schwitters (*supra* fn. 83), 260–261; W.J.P.M. Fase (*supra* fn. 95), 52.
[99] Kamerstukken I, 1899–1900, p. 288 and 290 (MvA).
[100] Kamerstukken II, 1896–1897, 159, no. 3, 12.
[101] Kamerstukken II, 1899–1900, 207, no. 6, 32. See also Kamerstukken II, 1899–1900, 159, no. 5, 24. This was also the main argument put forward by the employers themselves: R.J.S. Schwitters (*supra* fn. 83), 279–282.
[102] Kamerstukken I, 1899–1900, p. 221.
[103] B. Barentsen (*supra* fn. 68), 91.
[104] Kamerstukken II, 1896–1897, 159, no. 3, 13.

compensation as the main objective of the Bill strengthened the perception that the employers would be faced with the consequences of that aspiration and that for them the regulation would only amount to an increased responsibility, higher expenses[105] and more state intervention. Employer oriented parliamentarians had thus to be convinced of the fact that the advantages the regulation held were not solely for the benefit of the employees.[106]

44   The immunity was also based on the consideration that it would be unfair to make the employers pay twice for the same risk.[107] Other arguments referred to efficiency (preventing expensive procedures) and the need to maintain social peace.[108] That this immunity would undermine the preventive effect of the regulation was contradicted by the government with reference to the legally fixed exceptions to the immunity-rule[109] and the preventive effect of the mandatory contribution, the employer's concern for his reputation, and his sense of honor.

45   The OW did not offer full compensation. The indemnity was limited to 70% of the average day income. This was also meant to have a preventive effect: It was believed necessary that workers 'had something to lose' in order to make them more precautious at work and to avoid deliberate injuries.[110]

46   As important as the arguments justifying the choice for the new regulation, are the arguments that were put forward to explain why certain alternative systems had not been chosen for.

47   A first possibility discussed among lawyers was an extension of the employer's responsibility based on the labour contract.[111] The position that the terms of the labour contract imply for the employer the obligation to keep his employees in good health, was especially defended by the Belgian lawyer Sainctelette[112] but found no adherents in the Netherlands. Some objected on principle against state interference with the content of private labour contracts. Others only considered a rule of conduct – consisting of taking care for a safe work environment – possible on this ground, but not a duty to compensate.

---

[105] The premiums were indeed only to be carried by the employer.
[106] This probably explains why the immunity also encompassed emotional damage, although this damage was not covered by the OW.
[107] Kamerstukken II, 1898–1899, no. 16, p. 9.
[108] B. Barentsen (*supra* fn. 68), 92; C.J.H. Jansen/C.J. Loonstra (*supra* fn. 93), 92; Kamerstukken I, 1899–1900, p. 288–291.
[109] Material damages and damages caused by a criminal offence by the employer; B. Barentsen (*supra* fn. 68), 91; Kamerstukken II, 1898–1899, no. 16, p. 9.
[110] Kamerstukken II, 1896–1897, 159, no. 3, 15; Kamerstukken II, 1897–1898, 182, no. 3, 12. This loss of full compensation for the employees also reciprocated for the abolition of the employers' civil liability (S. Klosse (*supra* fn. 16), 62).
[111] R.J.S. Schwitters (*supra* fn. 83), 252–254.
[112] He saw this responsibility as arising directly from the authority the employer could exercise over his employees, which made him accountable for the consequences of his orders, for example the damage in case of an industrial accident. Cf. *infra* no. 133.

There was also a practical objection: When concluding the contract, the employer could always exonerate for this responsibility.

Another option was the widening of the employer's civil liability. One way of 48
doing this was by turning around the burden of proof. The employer would
then be held responsible unless he could prove the contrary. The government
refuted this solution because it would only shift the hardness from the employees to the employers. The rejection of civil liability as a means to allocate adequately the damages of work-related injuries appeared to be general.[113]

A first element in the expressed criticism was the argument that an expanded 49
civil liability could not guarantee compensation for the employee and would
for example offer no protection from possible insolvency of the employer.[114]
The nature of the compensation was considered inappropriate as well. Inappropriate for the victim, because in a tort action damages are usually awarded
in a lump sum. Regardless of the question whether a lump sum can ever sufficiently compensate for lost income, awarding a lump sum presupposes that the
victim is able to manage it effectively. If that appears not to be the case, the
employee-victim will have to fall back on poor relief again. But there is also a
disadvantage for the employer since the amount of the damages under tort law
is not fixed.[115]

Another argument was the unrealness of the freedom of contract for the em- 50
ployee. One was aware of the fact that the employee could hardly be considered a really equal contract party to the employer. This meant that employees
would probably renounce a possible claim and end up at charge of the community all the same. And if they decided to take action, this could mean a
threat for the social peace, taking into account the conflictuous character of
the civil procedure.[116] The bad experiences abroad at this point, especially in
Germany, strengthened this conviction.[117]

Schwitters also suggests that the choice against an expansion of civil liability, 51
represented in a way a form of legal conservatism. The preference for a compensation structure under public law would have been an attempt to protect the
dogmas of civil law against an amelioration of the legal position of the industrial injured that appeared to be inevitable.[118] Employers would also have had
an interest in getting rid of the civil procedure because it too strongly indicated their individual responsibility. The OW was in that perspective a way of rationalizing the responsibility for industrial injuries because it emphasized that

---

[113] R.J.S. Schwitters (*supra* fn. 83), 256.
[114] B. Barentsen (*supra* fn. 68), 74–75; R.J.S. Schwitters (*supra* fn. 83), 255.
[115] R.J.S. Schwitters (*supra* fn. 83), 256.
[116] One was indeed afraid that this could lead to social and socialist agitation. R.J.S. Schwitters (*supra* fn. 83), 185 and 264.
[117] B. Barentsen (*supra* fn. 68), 75; R.J.S. Schwitters (*supra* fn. 83), 244; Kamerstukken I, 1899–1900, p. 289.
[118] R.J.S. Schwitters (*supra* fn. 83), 245.

people bore responsibility not because they were sinful or free, but because they were part of a collectivity to which certain costs were imputed in order to solve a social problem.[119]

3.2.2 The Second Shift

52   In 1967 the *Wet op de Arbeidsongeschiktheidsverzekering* (Disability Insurance Act, WAO) came into force. This new act not only replaced the OW, but also integrated the *Invaliditeitswet* of 1919 (Invalidity Act, IW). These legislative changes were part of a new phase in the history of social insurance and social security that started in the postwar era. Due to the misery of World War II and the economic collapse of the 1930s a general belief had grown in Western Europe that the community had a responsibility to assume for the well being of its members.[120] In the Netherlands this opinion was articulated by the so-called Van Rhijn-committee that was entrusted with determining general guidelines for the future development of the social insurance. In its first report (1945) the committee stated that the system of social insurance had to involve the public at large and it therefore proposed a mixed system of social insurance and welfare. This meant that solidarity would come to the fore to the prejudice of the insurance notion.[121] This new ambition – that ultimately would lead to the modern welfare state – could also be realized financially thanks to a favourable economic climate.[122]

53   Against this background it does not come as a surprise that in comparison with the introduction of the OW the institution of the WAO did not provoke much discussion or objections. When the OW was proposed, the idea of a 'night watchman state' was still very strong. As a result the interference of the state, which was implied by that Act, had to be profoundly justified.[123] The discussions concerning the legal foundation of the Act were much more emotionally charged than they were 70 years later. The WAO was indeed adopted with a striking consensus.[124] This new Act was seen as the final piece of the social insurance building, a piece of legislation that marked a new era in social insurance and commanded admiration for both its magnitude and vision.[125] And with regard to its unique position in the world (cf. infra), there was no doubt but pride.[126]

---

[119] R.J.S. Schwitters (*supra* fn. 83), 4. These reasons probably also explain the by Schwitters observed paradox that civil action became a closed road at the very moment that employees – due to the changing labour relations – could more easily bring themselves to undertake such action. R.J.S. Schwitters (*supra* fn. 83), 183.

[120] C.J.H. Jansen/C.J. Loonstra (*supra* fn. 93), 91 and 97.

[121] S. Klosse (*supra* fn. 16), 192.

[122] A. de Swaan, *Zorg en de staat. Welzijn, onderwijs en gezondheidszorg in Europa en de Verenigde Staten in de nieuwe tijd* (1996), 223.

[123] W.J.P.M. Fase (*supra* fn. 95), 50.

[124] B. Barentsen (*supra* fn. 68), 180; A.Ph.C.M. Jaspers, De politiek en de sociale verzekering in: A.Ph.C.M. Jaspers et al. (eds.), *De gemeenschap is aansprakelijk ... Honderd jaar sociale verzekering 1901–2001* (2001), 36; A. de Swaan (*supra* fn. 122), 223.

[125] G.P. Sijses, *Wet op de arbeidsongeschiktheidsverzekering* (1965), 12 and 15; Kamerstukken II, 1963–1964, 7171, no. 14, 1.

[126] A.Ph.C.M. Jaspers (*supra* fn. 124), 36.

The reason for enacting a new law seems again to have been the wish for bet- 54
ter compensation. However, this wish was not primarily aimed at the OW,
which was considered to be 'a fairly satisfying arrangement'[127], but at the IW,
the insurance that covered not work-related disability. The coverage of this latter Act was very poor, especially when compared to that of the OW.[128] In accordance with the abovementioned opinion expressed by the Van Rhijn-committee this sharp contrast was not judged justifiable any longer. Urgent
measures directed at an amelioration of the position of this category of disabled had thus to be taken.

Veldkamp, the Minister of Social Services and Employment who proposed the 55
new Bill to parliament and who can be considered as the spiritual father of the
WAO, pursued yet another goal. Veldkamp wanted to liberate social insurance
from the individualistic characteristics borrowed from private insurance.[129] As he
argued in his doctoral thesis, social insurance differs from private insurance in
that the first is based on the idea that common risks should also be carried commonly while the second is based on the *do ut* des-principle. With regard to social
insurances solidarity appears more determining than a fair balance in individual
exchanges.[130] Veldkamp wanted the new Act to be in line with this vision. In
practice this meant that the WAO would loose some insurance characteristics in
comparison with the former OW and the IW, such as the risk-relatedness of the
premiums and the connection between the amount of paid premiums and the entitlement to benefits.[131] The second objective then equally comes down to an extension of the compensation function of the WAO, because what we are facing
here is a lowering of the barriers to compensation (cf. *supra* no. 4).

According to Veldkamp one of the important individualistic features which 56
social insurance had to be purged from was the *risque professionnel*-ap-
proach.[132] The question whether a separate insurance based on this *risque professionnel* was still necessary, was widely discussed during the preparation of
a new IW. Veldkamp himself already responded this question negatively in his
doctoral thesis of 1949[133] and in some earlier articles.[134] The Van Rhijn-committee on the contrary wanted to maintain for the time being the more favourable
compensation conditions for the working since 'the employees had not entirely
accepted the risk involved with their job by their own free will'.[135] A few years
later, in 1948, the hesitations in the second Van Rhijn-committee had already

---

[127] Kamerstukken II, 1962–1963, 7171, no. 3, 1 and 5. However, also here the government saw
   room for improvement, by making the allowances inflation-proof for example (*ibid.*, 1).
[128] B. Barentsen (*supra* fn. 68), 181 and 183–184.
[129] Kamerstukken II, 1962–1963, 7171, no. 3, 1.
[130] G. Veldkamp, *Individualistische trekken in de Nederlandse sociale arbeidsverzekering: een
   critisch onderzoek naar de grondslagen der sociale arbeidsverzekering* (1949), 12–13.
[131] B. Barentsen (*supra* fn. 68), 185 *et seq.*
[132] G. Veldkamp (*supra* fn. 130), 181.
[133] G. Veldkamp (*supra* fn. 130), 159–182.
[134] *E.g.* G. Veldkamp, Enkele critische opmerkingen over het risque professionnel, [1946] *Sociaal
   Maandblad* (SM), 125–134.
[135] S. Klosse (*supra* fn. 16), 184–185 (my translation).

become stronger. One could not deny the advantages of the *risque social*-approach: no more causality problems, no more differences between the benefits based on the IW and those base on the OW, and an easier administration of the regulation. Still, one kept supporting a separate *risque professionnel* based arrangement. Arguments for this position were the financial implications of the *risque social*-solution and the lack of foreign examples.[136]

57   In 1957 however, a recommendation of the *Sociaal Economische Raad* (Social Economic Council, SER)[137] takes for the first time the future unification of the different disability insurances as a starting point,[138] and a year later the *Sociale Verzekeringsraad* (Social Insurance Council, SVR) shows itself in favour of one unified disability insurance based on the *risque social*. The SER-recommendation of 1960 would confirm this view.

58   Two arguments are put forward for this position. First argument is the position that the *risque professionnel* is superseded as a legal foundation for the OW and should be replaced by the job itself, the actual labour relation as the legal ground for a compulsory insurance.[139] This is a reference to the so-called 'theory of the just or postponed wage' that holds that the pay employees get must be sufficient to cover a loss of income caused by the occurrence of an external risk. A second argument is found in the fact that the OW since the important statutory change of 1921 had come to protect for damage that clearly went beyond the *risque professionnel*, such as traffic accidents and accidents during company excursions.[140] Also Veldkamp had indicated the 1921-change as a rapprochement between the legal foundations of the OW and the IW, since the first was since then 'stronger oriented at the citizen outside the production process'.[141] In general one could make mention of a 'socializing' of the insured risk, an evolution away from the work-relatedness.[142]

59   The eventual Bill introduced in Parliament by Veldkamp was fully in line with these recommendations. As a consequence the explanatory memorandum perfectly reflects the views of Minister Veldkamp and the arguments put forward in the abovementioned recommendations. The central point of the memoran-

---

[136] P.S. Fluit/A.C.J.M. Wilthagen (*supra* fn. 97), 115–116. Especially the first reason would have been decisive: B. Barentsen (*supra* fn. 68), 96.
[137] This is a tripartite advisory council composed of representatives of employers, employees and independent experts.
[138] P.S. Fluit/A.C.J.M. Wilthagen (*supra* fn. 97), 116; S. Klosse (*supra* fn. 16), 195.
[139] See also G. Veldkamp (*supra* fn. 130), 67.
[140] P.S. Fluit/A.C.J.M. Wilthagen (*supra* fn. 97), 116; S. Klosse (*supra* fn. 16), 201–202. The definition of 'industrial accident' formulated in article 1 had changed to all accidents 'occurred to the employee concerning his employment' (my translation).
[141] G. Veldkamp, [1946] SM, 126 (my translation); G. Veldkamp (*supra* fn. 130), 96–98.
[142] B. Barentsen (*supra* fn. 68), 95 *et seq.* Klosse mentions that already in 1922 a bill was proposed that can be seen as an attempt to move from causality to finality or, in other words, to (partially) introduce the *risque social*. The Bill stipulated that disability caused by an industrial accident would be considered as caused by sickness during the first 6 weeks (S. Klosse (*supra* fn. 16), 108–109).

dum is the perspective of the *risque social*: decisive is not how someone got unable to work, but the fact that he or she is unable to work and that this disability has social consequences.[143] The starting point of the arrangement is no longer causality (what is the cause of the medical problem that led to disability?) but finality (is there a medical problem that led to disability and thus a loss of income?).[144] As mentioned earlier, it was not so much that the *risque professionnel* was a very unsatisfying system, but the aspiration to adequately compensate the non-working disabled that constituted the strongest motive underlying the new Act.[145]

As a justification the memorandum referred to the 'theory of the just wage' (cf. *supra* no. 58) that by then had been generally accepted as the legal foundation for most social insurances in favour of employees (such as the IW). The view was that this theory was not an original but a merely derived principle. Social justice demanded that every disabled man should have a scope to develop himself regardless of the cause of his disability. In other words: The actual basic principles were the right to self-development and the right to equal opportunities. Since an individual could not be expected to realize these rights for himself, a compulsory social insurance that would guarantee an entitlement to reasonable benefits irrespective of the cause of disability, could contribute greatly to the realization of these principles.[146] The principle of equality thus seems to be an important element in the justification: disabled persons must have equal rights (to self-development) as not disabled, and no discrimination is allowed between the disabled.[147]   60

Yet, the introduction of the *risque social* approach is more than a mere equalization of work and non-work related injuries. It also expresses that the occurrence of work-related injuries cannot be considered any longer as a private risk. The accident risk is considered a social risk, inherent to our production method. It is a risk that is not primarily caused by the pursuit of profit of companies, but by the satisfaction of our social needs that demands for an ever increasing and more perfectionized production system. The burden of this risk should then be equally divided over all companies.[148] An argument that was used under the OW is thus recuperated but now linked to an adverse conclusion. Where the qualification of the accident risk as inherent to the production system in 1901 led to the allocation of this risk to the employer, the conclusion now reads that it concerns a social risk chosen (and thus to be born) by society.[149]   61

---

[143] Kamerstukken II, 1962–1963, 7171, no. 3, 2.
[144] B. Barentsen (*supra* fn. 68), 182.
[145] B. Barentsen (*supra* fn. 68), 183–184.
[146] Kamerstukken II, 1962–1963, 7171, no. 3, 2.
[147] P.S. Fluit/A.C.J.M. Wilthagen (*supra* fn. 97), 118.
[148] Kamerstukken II, 1962–63, 7171, no. 3, p. 2–3 (MvT); G. Veldkamp, [1946] SM, 132–134. Veldkamp also argues that the risk-related premiums of the *risque professionnel* approach interfere with a just sytem of prices and incomes (*ibid.*, 126 and 133).
[149] P.S. Fluit/A.C.J.M. Wilthagen (*supra* fn. 97), 118–119. In contrast with the OW the WAO-premiums are equally to be paid for by employers and employees.

This changed risk perception probably demonstrates the short amount of time that was needed for society to get used to and to accept the industrial production method that only originated a century before.

62   In both Houses of Parliament the proposed Bill was well received. In fact, mention was made of an important landmark in the historical development of social insurance.[150] Both Houses also gave their unqualified assent to the *risque social* foundation.[151] Still some reservation was made by the *Tweede Kamer* (Lower Chamber, TK) with regard to the generous character of the offered compensation. In order to prevent fraud a minimum disability of 25 instead of 15 percent was esteemed necessary. Nor did one want the benefit to get the character of a compensation for moral harm.[152] The TK also asked[153] to provide the possibility to refuse a benefit in case the disability was the result of strongly condemnable acts committed by the victim himself, but the minister doubted the preventive effect of such a possibility and saw this kind of general prevention as a task of the criminal legislator.[154]

63   At the other side the TK formulated no objection concerning the reintroduction[155] of the employers' civil liability. The EK however, did. She considered it too strong an interruption of the continuity of the insurance.[156] The reintroduction of the employers' civil liability had been put forward in a recommendation of 1963 of the SVR. The majority of this council reasoned that at this point the new Act should seek connection with the arrangement in the *Ziektewet* of 1913 (Sickness Benefits Act, ZW 1913) and not the former OW taking into account the similarity of the character of the insured risk, being the general risk of not being able to earn an income through employment by sickness or accident. The Minister[157] and TK[158] followed this line of reasoning. The argument that the social peace in companies would be endangered by (an increase of) civil lawsuits, was not accepted. First, one expected these lawsuits to be dealt with in a businesslike way since the proceedings would generally be oriented against the employer's insurer. Secondly, trade unions were supposed to guard against irresponsible claims by employees. And finally, social peace would not be served by the shortcutting of civil rights of employees. This last argument was a reference to the compensation objective. The basic assumption appeared to be that the new Act could not be allowed anymore (as

---

[150] Kamerstukken II, 1964–1965, 7171, no. 14, 1. This opinion was however not shared by the EK that on the contrary expressed its disappointment with regard to the limited scope of the Bill and the lack of 'a more modern methodology on the subject of the allocation of social justice' (my translation). Kamerstukken I, 1965–1966, 7171, no. 23, 1.
[151] Kamerstukken II, 1964–1965, 7171, no. 14, 2–3; Kamerstukken I, 1965–1966, 7171, no. 23, 1.
[152] Kamerstukken II, 1964–1965, 7171, no. 14, 3.
[153] Kamerstukken II, 1964–1965, 7171, no. 14, 3.
[154] Kamerstukken II, 1964–1965, 7171, no. 15, 7–8.
[155] The OW of 1901 rendered the employers immune against this liability.
[156] Kamerstukken I, 1965–1966, 7171, no. 23, 4.
[157] Kamerstukken II, 1962–1963, 7171, no. 3, 19–20.
[158] Kamerstukken II, 1964–1965, 7171, no. 14, 4.

did the OW) to hinder the employee to receive full compensation.[159] A minority in the council argued that the employers' civil liability should be excluded except in cases where this would lead to unfairness, for example in cases of criminal offence. A limitation on the employees' civil rights of that kind seemed justified taking into account the character of the right they could exercise under the WAO, i.e. a partial though inflation-proof benefit even in case of contributory negligence. Reference was also made to similar arrangements in other countries.[160]

Another matter in dispute between the Minister and the parliament was the right of recourse. The Bill provided a right of recourse towards the party that could be held responsible according to civil law. A right of recourse towards the employer of the victim however, was only possible in case of foul play or deliberate recklessness.[161] The TK objected to a general right of recourse on practical grounds; she expected long and expensive proceedings with little result in the end.[162] The Minister however replied that he trusted on the wisdom of the administration to decide when it was useful to exercise the right of recourse.[163]   64

The introduction of flat rate premiums on the other hand was no point of difference. The SVR and the SER had ultimately advised to make the premiums risk-related only for the benefits of the first two years of disability, because only then the continuing of a relationship with the former industry sector could be assumed. Minister Veldkamp rejected this proposition and considered even the smallest risk-relatedness of premiums inconsistent with the peculiar character of the social insurance.[164] The TK agreed with this view stating that since business sectors could hardly influence the extent of the risk any longer (due to the *risque social* foundation of the WAO), no motives seemed to be left to argue for risk-related premiums.[165]   65

The EK was mainly worried about the amount of the premiums. It feared that the burden on trade and industry could become to heavy, especially in comparison with foreign companies.[166] A counter-argument, which was not invoked but was pre-emptively parried by Minister Veldkamp in the explanatory memorandum, was the preventive effect of risk-related premiums that would go lost in the WAO.[167] Veldkamp seems to admit this but points at the fact that research has shown that the indirect cost of work-related injuries is higher than the direct cost and that by consequence an incentive for employers to promote   66

---

[159] The compensation for immaterial damage was in particular mentioned. Kamerstukken II, 1962–1963, 7171, no. 3, 19–20; Kamerstukken I, 1965–1966, 7171, no. 23a, 8.
[160] Kamerstukken II, 1962–1963, 7171, no. 3, 19.
[161] Kamerstukken II, 1962–1963, 7171, no. 3, 20–21.
[162] Kamerstukken II, 1964–1965, 7171, no. 14, 4.
[163] Kamerstukken II, 1964–1965, 7171, no. 15, 10–11.
[164] Kamerstukken II, 1962–1963, 7171, no. 3, 30–32.
[165] Kamerstukken II, 1964–1965, 7171, no. 14, 9.
[166] Kamerstukken I, 1965–1966, 7171, no. 23, 2 and 5.
[167] Kamerstukken II, 1962–1963, 7171, no. 3, 32.

safety will remain present in the WAO.[168] Veldkamp was moreover convinced that there were alternative solutions that could have a similar effect, such as an intensified labour inspection with severe sanctions.[169]

3.2.3 A Third Shift?

67   What gloriously had been launched as the nation's pride in the area of social security started to turn into a national nightmare only ten years later. Where the expectation was that some 200.000 people would have to use the WAO, already in 1976 the figure of half a million was passed. The effect of this 'success' of the compensation scheme on the national budget became worrying, especially when the Netherlands – in view of the depression of the 70s – was faced with the necessity of an austerity policy. A first economy measure was taken in 1985 when the maximum benefit was lowered from 80 to 70%. Two years later the disability benefit scheme was also cleared from the risk of unemployment.[170]

68   Still, the number of disabled people kept increasing and approached the magic boundary of 1 million[171] people, which brought prime minister Lubbers to the legendary phrase: *'Nederland is ziek'* (the Netherlands is sick).[172] In 1993 a parliamentary research committee under the chairmanship of Buurmeijer made public the conclusions of an analysis that already had been clear for some time: Employers and trade unions had for years unitedly misused the WAO as a gentle way to lay off redundant staff so that the WAO constituted in practice 'a vessel of hidden unemployment'.[173] New measures were not long in coming. The benefits for new disabled people were made dependent on age and would thus vary in time and size; the criteria for assessing disablement were strengthened; and employers were confronted with premium differentiation and a no-claims bonus system.[174] Also the organisational structure of the WAO was modified: Since employers and trade unions had proved to behave irresponsible, a public body, the *Landelijk Instituut Sociale Verzekeringen* (National Institute Social Insurances, Lisv) took over the tasks of the industrial insurance board.

---

[168] Kamerstukken II, 1962–1963, 7171, no. 3, 32; G. Veldkamp (*supra* fn. 130), 177 (fn. 1).

[169] G. Veldkamp, [1946] SM, 134 and G. Veldkamp (*supra* fn. 130), 170.

[170] Before, people who were partially disabled and partially unemployed nevertheless received a full disability benefit.

[171] Indeed if the Netherlands were to have relatively as much disabled people as Germany, the figure would have had to be three times lower.

[172] Speech for the University of Nijmegen on 3 September 1990 (my translation)

[173] For the employers this way of acting had the advantage of limiting the costs of a social plan without upsetting the social peace. The employees for their part were assured of a relatively high income (originally 80% of their last wage) up to their retirement and escaped the stigma of unemployment. The benefits of the *Werkloosheidswet* (Unemployment Insurance Act, WW) were indeed lower and limited in time.

[174] Cf. the *Wet terugdringing arbeidsongeschiktheidsvolume* (Act Reducing Disability Rate, TAV) of 1992 and the *Wet terugdringing beroep op de arbeidsongeschiktheidsregelingen*, (Act Reducing the Demand for Disability Benefits, TBA) of 1993.

The first cabinet presided by prime minister Kok took down in the coalition 69
agreement a drastic reduction of the number of disabled and decided to take
measures that in a more direct way would confront the employers with the
costs of sick and disabled employees.[175] From 1996 on employers were obligated to remain paying 70% of the wage of a sick employee during the first
year.[176] This came down to an almost complete de facto privatization of the
Ziektewet of 1929 (Sickness Benefits Act, ZW 1929).[177] Two years later the
*Wet premiedifferentiatie en marktwerking bij arbeidsongeschiktheidsverzek-
eringen* (Disability Insurance (Differentiation in Contributions and Market
Forces) Act, Pemba) was introduced. This Act put the contribution for the
WAO entirely on the employers and prescribed that the premium would increase in function of the individual accident rate.[178] However, the option was
provided for the employer to be his own risk bearer. He could take out private
insurance or pay the disablement benefits himself during five years. The introduction of the Pemba implied to a certain extent a reversion to the situation
under the OW of 1901, where the risk of long-term disability was a full business risk. At the same time measures were taken to promote the engagement
of disabled by employers through the promulgation of the *Wet Reïntegratie
Arbeidsgehandicapten* (Disablement Rehabilitation Act, Rea).

After the number of disabled had gone down for a time[179], it restarted to in- 70
crease and new measures were envisaged. In June 2000 an advisory committee of experts under the chairmanship of Donner was installed and the *Wet
verbetering Poortwachter* (Gatekeeper Improvement Act, WVP) enacted. Research had shown that too less attention was paid with regard to preventing
disability and that, in the first year of disability, measures to get people back to
work were taken too late. In conformity with the coalition agreement of
1998[180] the new Act stressed the importance of prevention and rehabilitation:
Clear standards were set to employers, employees and *Arbodiensten* (Occupational safety and health services) with regard to the minimal required actions
to be taken in order to reach rehabilitation in a suitable employment.[181] Also
Arboconvenanten (covenants with regard to the working conditions) would be

---

[175] The coalition agreement of 1994 stated that the necessity was felt to come to a new balance
with regard to the division of tasks between government, citizens, employers and employees.
With regard to the WAO this meant that the financial risk had to be taken by those who were
responsible for a correct implementation of the scheme. It was moreover believed that the
introduction of premium differentiation and market competition would improve this implementation, result in more prevention and rehabilitation and would finally also reduce the disability volume. *Regeerakkoord* (Coalition Agreement) 1994, 4 and 16.
[176] The *Wet Uitbreiding Loondoorbetalingsplicht bij Ziekte* (Extension of Compulsory Sick Pay
Act, Wulbz) changed article 7:629 *Burgerlijk Wetboek* (Dutch Civil Code, BW) in that sense.
[177] A lot of employers took out private insurance to cover this new risk.
[178] The contribution consisted of a fixed basic premium and a differentiated extra premium.
[179] From 14,7% of the labour force in 1994 to 12,1% in 1997.
[180] *Regeerakkoord* (Coalition Agreement) 1998, 33–34.
[181] A lack of efforts thereto will be sanctioned with discontinuation of salary payment or even dis-
missal for the employee (art. 7:629, 3° BW and art. 7:670b, 3° BW) and with the extension of
the obligation to keep paying the salary for the employer (art. 71a WAO).

concluded for each industrial branche between employers, employees and the government in which engagements would be made on how to reduce safety and health risks. As an incentive to prevent accidents or deterioration of personal injuries, the obligation for employers to continue to pay their disabled employees was also made more severe by extending this obligation to a period of two years instead of one.[182]

71   The Committee-Donner delivered its report[183] in 2001 and suggested a profound reform of the scheme by which the WAO would be reserved for the fully and permanent disabled. This advice was also supported by the SER.[184] After long negotiations between the different interested parties, a compromise was reached that led to the replacement of the WAO by a new act, the *Wet Werk en inkomen naar arbeidsvermogen* (Work and Income According to Labour Capacity Act, WIA) that came into force on 1 January 2006. A new act was judged necessary and urgent by the government because it expected the pressure on the compensation scheme to grow due to the increase[185] and ageing of the labour force.[186] The government also wanted the scheme to become more 'activating', stressing what disabled people are still capable of doing instead of focusing on their disabilities. This means giving priority to the reintegration on the labour market over providing income-support.[187]

72   The WIA comprises of two schemes: the *Regeling Inkomensvoorziening Volledig Arbeidsongeschikten* (Fully Disabled Income Protection Regulation, IVA) for the fully and permanent disabled, and the *Regeling werkhervatting gedeeltelijk arbeidsgeschikten* (Partially Disabled Re-employment Regulation, WGA) for the partially disabled. The first scheme accords a benefit of 70% of the last earned salary[188] to people who have lost at least 80% of their earning capacity and this up to the retirement age.[189] Disabled people who have lost more than 65% of their earning capacity (but less than 80%) will re-

---

[182] It was moreover forbidden to supplement this benefit up to 100% of the salary by means of a collective labour agreement, as had happened with the original obligation concerning the first year.

[183] *Werk maken van Arbeidsgeschiktheid* (Putting work in labour capacity), 30 May 2001.

[184] *Werken aan arbeidsgeschiktheid* (Investing in labour capacity), 22 March 2002.

[185] This increase is expected to be mainly the result of a higher participation of women and older people, who represent high-risk groups with regard to work-related injuries.

[186] Note in response to the parliamentary report, 26 May 2005, 5; Kamerstukken II, 2004–2005, 30 034, no. 3, 4.

[187] Kamerstukken II, 2004–2005, 30 034, no. 3, 3. This all fits in a broader line of policy that tries to reduce avoidable and unintended use of social security by stimulating employers and employees by means of financial incentives to take their own responsibility and positively influence their own risks (Kamerstukken II, 2004–2005, 30 034, no. 3, 24).

[188] A retroactive increase up to 75% is envisaged when the rise of fully and permanent disabled in 2006 has not exceeded the number of 25000. In that case the premium differentiation of the Pemba would also be abolished. Kamerstukken II, 2004–2005, 30 034, no. 3, 22, 58 and 78 *et seq.*

[189] The obligation to continue paying 70% of the last wage during the first two years of disability remained intact under this new act. The IVA-benefit consequentially starts from the third year on.

ceive a benefit on basis of the WGA. The benefit offered by this scheme depends on a qualifying period and evolves after a certain time and in function of the individual employment record to a non-earnings related benefit (percentage of the minimum wage). It is organised in such a way as to give financial incentives to the disabled to use his remaining labour capacity. Financial incentives are also provided towards the employers: Keeping or getting a partially disabled at work is awarded with a no risk policy[190] and a reduction on the social insurance contribution. Employers have the choice to carry the risk of partially disablement themselves[191] or to have it covered through private or public insurance.[192]

Initially the government considered supplementing the new Act with an extra provision, the *Extra Garantieregeling Beroepsrisico's* (Occupational Risks Additional Income Protection Guarantee, EGB) for the victims of work-related injuries and occupational diseases, i.e. the pure *risque professionnel*, by means of the reintroduction of a mandatory private insurance.[193] This would indeed mean a partial shift back to the situation in 1901.[194] In its request for an advice from the SER[195] the government saw two reasons for this. First of all, the government feared that the income-support provided for the partially disabled under the WGA might not be in accordance with the ILO-convention No. 121 of 1964 concerning benefits in the case of occupational injury[196], nor with Part VI of the European Code on Social Security.[197] Secondly, the government expected that the considerable decline in income (especially for the higher incomes) in case of partially disablement under the new WGA-regime would provoke more civil proceedings on the basis of the employer's liability to offer a safe working place, and bring about a 'claim culture'. 73

The SER responded[198] to the government not to worry about the first reason, but admitted that the developments with regard to civil proceedings, the employers' liability and the possibility of private insurance should be followed closely in the near future. It also suggested that since both employers and employees are already facing long and uncertain proceedings, the introduction of a *risque professionnel*-insurance combined with the employers' civil immuni- 74

---

[190] In case this employee would fall sick again, the employer will be exempted from the obligation to continue paying him 70% of his last wage. Nor will this employee be taken into account when calculating the premium differentiation.

[191] This is supposed to be a possible incentive for rehabilitation efforts.

[192] In this latter case, the incentive towards prevention and rehabilitation is supposed to follow from the premium differentiation, which is remained with regard to the WGA-benefit. Kamerstukken II, 2004–2005, 30 034, no. 3, 83.

[193] The proposed characteristics of the insurance conditions strongly resemble these of the current Belgian scheme.

[194] Cf. *supra* no. 32 *et seq.*

[195] Request from minister De Geus dd 19 November 2003.

[196] See also S. Klosse, Van WAO naar WIA: een verantwoorde omslag?, [2005] *Sociaal Recht* (SR), 247–257.

[197] In the first case the requested minimum standards might not be reached; in the second case the employee's contribution to the medical costs seems problematic.

[198] *Verdere uitwerking WAO-beleid* (Further elaboration disablement policy), 20 February 2004.

ty, could indeed have benefits for all interested parties. The trade unions are holding the same opinion: They believe that the extension of the WIA with such a kind of insurance would imply to combine the best of two worlds, and would probably also be conducive to prevention.[199] Whether this extra provision will ultimately become reality or not, remains to be seen.[200]

3.3   The Shifts in England

3.3.1   A First Shift?

75   Also in Britain the first substantial policy shift in compensating for work-related injuries can be situated around the turn of the 19th to the 20th century, although the matter had been an element of discussion since the first half of the 19th century. The introduction of the *Workmen's Compensation Act* of 1897 (WCA)[201] represented a revolution in juridical thought, 'an entirely new doctrine'.[202] According to the so-called Digby Committee it was '... difficult to overrate the boldness and importance of the step then taken by the legislature'.[203] Still, it seems that in order to be able to take this step the English first needed a backward move by way of run up. After all, at a moment when workmen most needed the protection available before 1897 by the Common Law, it was interpreted by the courts in a manner greatly to their detriment.[204] Because of this long run up it seems necessary to describe the different motives and counter-arguments for this first shift in a more historical evolutionary way.

76   The right of compensation afforded by the Common Law offered a rather meagre protection, which was however hardly felt in practice before the rise of modern industry to whose requirements, it would prove to be inadequate. The establishment of the personal responsibility of the employer, which was required for a successful action based on tort, had become more difficult than ever before and gave rise to legal proceedings causing costs and delays.[205] For an employee, going to court implied anyhow risking to be dismissed with the certainty that no other master in the district would hire the 'troublemaker' again.[206] Trade unions had no regular legal advisers that could assist the victims[207] and the social composition of courts further limited the chances of success.[208] Moreover,

---

[199] Press release *Federatie Nederlandse Vakbeweging* (Dutch Trades Union Federation, FNV) dd 23 February 2005.
[200] The government is currently still waiting a response to its request (of 11 March 2005) for an informal opinion of the ILO with regard to the ILO-proofness of the new Act.
[201] The Act came into force 1 July 1898.
[202] Holman Gregory Report, p. 7.
[203] Report of the Departmental Committee on Compensation for Injuries to workmen, 1904, p. 11.
[204] J.C. Brown, *Disability Income. Part 1. Work-related injuries* (1982), 2; A. Wilson/H. Levy, *Workmen's Compensation, vol. 1: Social and Political Development* (1939), 24.
[205] A. Wilson/H. Levy (*supra* fn. 204), 5–6.
[206] P.W.J. Bartrip, *Workmen's Compensation in Twentieth Century Britain. Law, History and Social Policy* (1987), 3.
[207] A. Wilson/H. Levy (*supra* fn. 204), 25.
[208] P.W.J. Bartrip (*supra* fn. 206), 3.

Common Law offered no remedy in case of fatal accidents[209] and provided no compensation for 'Acts of God'. And when a legal claim was directed against a fellow servant, this latter generally disposed of no means to satisfy the judgement against him.[210]

As already mentioned these shortcomings of the tort system were aggravated   77
by a judicial interpretation that narrowed the possibilities for compensation through the imposition of three legal fictions that were based on economic concepts and assumptions that in no way conformed to reality.[211] The doctrines of common employment[212], *volenti non fit iniuria* and contributory negligence served as strong additional defences to the employers and rendered the acquisition of compensation even more difficult and illusionary.[213] This constituted undoubtedly the particular result aimed at by the judges.[214] In *Priestley v. Fowler* (1837), the case that launched the common employment doctrine holding that an employer is not vicariously liable to one of his employees for an injury occasioned by the negligence of another, public policy rather than legal considerations informed the decision as Lord Abinger stated that to decide the other way would be to 'open the floodgates' of litigation with 'alarming' consequences.[215]

In short, by the middle of the 19th century the Common Law had come in   78
practice to afford little protection to the workman. Attempts to bring about a change in this situation were in general not primarily motivated by an endeavour to obtain damages as by a concern for safer working conditions. It was believed that safe systems of work would be most effectively achieved if accidents were made expensive to employers. This 'expensive accident theory' was also present in the proposal of the Chadwick Commission (1833)[216] that

---

[209] This changed when the Fatal Accidents Act of 1846 (Lord Campbell's Act) came into force. This Act, that not only applied to workmen's compensation, recognised an action for the dependants where the deceased, had he lived, would have had one: P. Cohen, *The British System of Social Insurance* (1932), 196–197; A. Wilson/H. Levy (*supra* fn. 204), 29–30.

[210] A. Wilson/H. Levy (*supra* fn. 204), 6; P. Cohen (*supra* fn. 209), 196.

[211] The defence of *volenti* in particular displays a preoccupation with contract and the conception of the employee as a 'free agent', able to protect himself either by extracting an express guarantee of his safety from his master, or else refusing to undertake risky work (W.R. Cornish/G. Clark, *Law and Society in England 1750–1950* (1989), 497–498; A. Wilson/H. Levy (*supra* fn. 204), 25–26). Another (economic) assumption consisted in the belief that an employee in agreeing the stipulated remuneration, had also accepted the risks associated with the work engaged in, including the risk of all fellow-servants' negligence (W.R. Cornish/G. Clark (*supra* fn. 211), 498–499; A. Wilson/H. Levy (*supra* fn. 204), 12).

[212] According to this doctrine the employee was deemed to have entered into an implied contract with the employer to accept the risks, which might arise from the carelessness of his fellow employees, whereas before, Common Law had placed upon the employer responsibility for injuries caused by negligence or breach of duty not only by the employer himself, but also by any person in his service when acting within the scope of his employment. J.C. Brown (*supra* fn. 204), 1–2.

[213] P.W.J. Bartrip (*supra* fn. 206), 6–7; P. Cohen (*supra* fn. 209), 196; W.R. Cornish/G. Clark (*supra* fn. 211), 508; A. Wilson/H. Levy (*supra* fn. 204), 24–27.

[214] W.R. Cornish/G. Clark (*supra* fn. 211), 497.

[215] As quoted in P.W.J. Bartrip (*supra* fn. 206), 6.

[216] This Royal Commission, chaired by Chadwick was installed after Lord Ashley had introduced a Bill into Parliament which included industrial injury compensation clauses, and was appointed to fully examine factory conditions throughout the country.

suggested making employers financially responsible for maintaining those injured in their service, thereby increasing their economic interest in safety. First
progress was however not made according to Chadwick's idea of safety via
economic deterrence, but by the introduction of Factory Acts (in 1833 and
1844) that laid down certain minimum standards of industrial safety. The compensation clauses of the latter Act were however very limited and of marginal
practical use.[217]

79   From the 1860s on labour propositions for reform were for a great deal directed
       at the abolishment or at least the limitation of the common employment doc-
       trine. The employers firmly objected to such an extension of their liability and
       showed themselves more in favour of voluntary insurance through mutual relief
       funds with contributions by both employer and employee. As the threat of a dis-
       advantageous change of liability law grew, they even suggested to have a system
       of general insurance examined, as insurance schemes were a good means of ce-
       menting good relations between masters and men, and of avoiding litigation.[218]

80   After the Trades Union Congress (TUC) (established in 1868) had come into
       action, the efforts aimed at limiting the common employment doctrine result-
       ed – under a Liberal government – in the enactment of the *Employers' Liabili-
       ty Act of 1880*[219] that allowed industrial manual workers to sue their employers
       vicariously for the negligence of non-manual superintendents and for certain
       other defaults. A legal argument was put forward to underpin the Act: Aban-
       doning the defense of common employment would mean getting rid of a de-
       liberately judge-made discrimination in the law and putting working people
       back on an equal footing with the rest of community.[220]

81   The counter-arguments were multiple. The Act was considered an inadmissi-
       ble interference with the right of property and the freedom of contract, and
       would induce a litigious atmosphere that would set class against class. Placing
       financial responsibility upon the employers for some accidents could more-
       over only be met by cuts in wages and would threaten English industry and
       leave the employers without resources to show paternal generosity to all.
       Moreover, the Act would render workers less careful and would take away a
       motive for thrift. After all, workmen, in dangerous employment at least,
       should insure themselves against accidents.[221] The Tories, in opposition at that
       time, were expected to defend the employers' interests but tried to embarrass
       the Liberal government by arguing that the Bill was insufficiently liberal.[222]

---

[217] P.W.J. Bartrip (*supra* fn. 206), 3–5.
[218] E.P. Hennock, British Social Reforms and German Precedents. *The Case of Social Insurance
      1880–1914* (1987), 45–46 and 49–50.
[219] 43 en 44 Vic. C42.
[220] P.W.J. Bartrip (*supra* fn. 206), 8; W.R. Cornish/G. Clark (*supra* fn. 211), 522; E.P. Hennock
      (*supra* fn. 218), 40.
[221] P.W.J. Bartrip (*supra* fn. 206), 8; W.R. Cornish/G. Clark (*supra* fn. 211), 521–522; A. Wilson/
      H. Levy (*supra* fn. 204), 39–40.
[222] A. Wilson/H. Levy (*supra* fn. 204), 39.

The eventual result of the Act for the workmen was rather limited. Although it appeared to have given an incentive to adopt safety precautions[223], it still did not offer a certain and easy means of obtaining compensation since most social, economic and legal barriers remained.[224] Nor could employees count on the sympathy of judges and lawyers who perceived the Act as unjust towards employers.[225] The Act also caused problems of interpretation and a stream of litigation.[226] But most importantly, there was the possibility of 'contracting out' for the employer, a practice that was declared legal by the Queen's Bench in the *Griffiths v. Earl of Dudley* case of 1882 and would become the most important thorn in the flesh of trade unions and their employees for the years to come.[227] The employers had equally reasons to be dissatisfied: Their legal liability was uncertain, the costs were high and they feared social frictions and irritations in the factories.[228]

82

Because of the general dissatisfaction with the Act and, in particular, with the litigation to which it gave rise, and due to the fact that Conservatives as well as Liberals competed for the working-class vote[229], further reforms were sought after. Eliminating the inadequacies of the fault liability system and not securing better compensation remained thereby the first concern of labour organisations. Their focus was at the prevention of accidents and they believed that safety could only be realised if responsibility would rest clearly and exclusively upon employers.[230] Logically then they mainly aimed at the abolition of contracting out and opposed to any system of insurance, which was considered as a way for employers to escape their liability.[231] The attitude of trade unions could thus be evaluated as not very conducive to more progressive measures.[232]

83

As remarkable however was the fact that the position of the judiciary was. Wilson and Levy refer to a letter of Lord Shand published in the *Times* of 24 August 1880 in which this Lord of Appeal states (as a reaction to the Employers' Liability Bill):

84

---

[223] A. Wilson/H. Levy (*supra* fn. 204), 47.

[224] W.R. Cornish/G. Clark (*supra* fn. 211), 525–526; P. Cohen (*supra* fn. 209), 197.

[225] W.R. Cornish/G. Clark (*supra* fn. 211), 526; E.P. Hennock (*supra* fn. 218), 42; A. Wilson/H. Levy (*supra* fn. 204), 42.

[226] A. Wilson/H. Levy (*supra* fn. 204), 46.

[227] W.R. Cornish/G. Clark (*supra* fn. 211), 526; A. Wilson/H. Levy (*supra* fn. 204), 47. Trade unions regarded contracting out as a means of avoiding the penalties for negligence the Act provided for, and thus as a curtailment of its possible preventive effect: E.P. Hennock (*supra* fn. 218), 48.

[228] A. Wilson/H. Levy (*supra* fn. 204), 47.

[229] W.R. Cornish/G. Clark (*supra* fn. 211), 528; A. Wilson/H. Levy (*supra* fn. 204), 56.

[230] J.C. Brown (*supra* fn. 204), 93; W.R. Cornish/G. Clark (*supra* fn. 211), 525; A. Wilson/H. Levy (*supra* fn. 204), 53.

[231] P.W.J. Bartrip (*supra* fn. 206), 9; W.R. Cornish/G. Clark (*supra* fn. 211), 513; A. Wilson/H. Levy (*supra* fn. 204), 54.

[232] A. Wilson/H. Levy (*supra* fn. 204), 53.

'The injustice and difficulties arising from either course of legislation would be avoided, and a complete remedy would be supplied for the evils now existing, by leaving the law of liability as it now stands, but providing by statute for a compulsory system of insurance. For every accident, including those caused by the thoughtless act of the sufferer, a fund would be at once available. The employer would be saved from the ruin which might overtake him by one serious occurrence, and the workman would have a remedy at hand without recourse to charity or the rates'.[233]

Legal conservatism thus brought the judiciary to plea in favour of compulsory insurance. This proposal was clearly inspired by the knowledge of the similar practice of the *Knappschaften* (miners' unions) in the Prussian mining districts.[234] Compulsory insurance however was not thought suitable to England, as being destructive of thrift and self-reliance[235] and as irreconcilable with the English nature.[236]

85   In 1893 the Liberal[237] government proposed a Bill based on the 'general principle' that masters were responsible for the acts of their servants. The Bill abolished the defence of common employment, limited the doctrine of *volenti* and prohibited contracts restricting the application of the Act of 1880. It also simplified the procedure by means of which a workman could seek his statutory remedy. Due to a difference with the House of Lords with regard to the prohibition[238] of contracting out, the government finally dropped the Bill. But meanwhile an amendment of Lord Chamberlain had indicated the lines of future reform:

'That no amendment of the law relating to Employers' Liability will be final or satisfactory which does not provide compensation to workmen for all injuries sustained in the ordinary course of their employment and not caused by their own acts or default'.[239]

Chamberlain's point was that if employers were to be held liable for accidents over which they had no personal control, it was fair and logical that all victims who were not themselves negligent, thus including those injured by an 'Act of God', should gain redress. He thus revolutionised industrial compensation by calling for insurance rather than a negligence basis.[240] But his proposal was vulnerable to an old Trades Union argument: a compulsory insurance would take away the incentive to prevention as it made accidents cheaper for the em-

---

[233] A. Wilson/H. Levy (*supra* fn. 204), 42.
[234] E.P. Hennock (*supra* fn. 218), 44–45.
[235] A. Wilson/H. Levy (*supra* fn. 204), 56.
[236] E.P. Hennock (*supra* fn. 218), 44.
[237] In 1891 the Liberal party had taken over the TUC proposals on employers' liability: E.P. Hennock (*supra* fn. 218), 50.
[238] Whereas the abolition of the defence of common employment created formal equality of rights between workmen and the general public, the proposal to prohibit contracting out would do the opposite: E.P. Hennock (*supra* fn. 218), 56.
[239] Quoted by P. Cohen (*supra* fn. 209), 198.
[240] P.W.J. Bartrip (*supra* fn. 206), 9.

ployer instead of more expensive.[241] Little by little however, Chamberlain would succeed in convincing the unions of the fact that beside the preventive action for damages based on negligence a curative entitlement to smaller sums should be placed.[242]

Thanks to trade union activity the question of a Compensation Act irrespective   86
of cause of accident, would become an important issue in the General Election of 1895[243] and in 1897 the Workmen's Compensation Act[244] would be enacted. The Act imposed upon employers in certain industries liability to give compensation in cases of 'personal injury by accident arising out of and in the course of employment of the workmen', independent of the question whether or not there had been negligence on the employers' part or of anyone employed by them. However, no compensation would have to be paid in case of 'serious and willful misconduct' of the employee.

As for the Netherlands this first shift can be described as a means to provide a   87
cheaper and more reliable legal remedy to victims of work-related injuries.[245] Saving the injured workers from destitution by facilitating some measure of income maintenance was surely one of the Act's main objectives.[246] As mentioned before, this was however not the only main objective. As important was improving industrial safety by making accidents more expensive to employers[247] as these were considered best placed to take preventive measures.[248] That this latter aim was at least of equal weight to the first, is reflected in two features of the new compensation regime.

First of all, the Act did not compel employers to insure – as insurance was seen   88
as disruptive of taking precautionary measures, which meant for the workmen accepting the risk of insolvency of their employer and thus lose on compensation security.[249] On the other hand, it was expected that most employers would insure in order to be able to budget their expenses with greater certainty.[250] The absence of compulsory insurance was furthermore related with ideas of self-reliance and freedom that also served as a justification for the preservation of the possibility of contracting out[251] and that were seen as inherent to

---

[241] A. Wilson/H. Levy (*supra* fn. 204), 60; E.P. Hennock (*supra* fn. 218), 63.

[242] W.R. Cornish/G. Clark (*supra* fn. 211), 528; E.P. Hennock (*supra* fn. 218), 53 *et seq.*

[243] P.W.J. Bartrip (*supra* fn. 206), 10; A. Wilson/H. Levy (*supra* fn. 204), 63.

[244] 60 and 61 Vic. Ch.37.209

[245] P.W.J. Bartrip (*supra* fn. 206), ix.

[246] P.W.J. Bartrip (*supra* fn. 206), 10; J.C. Brown (*supra* fn. 204), 94; A. Wilson/H. Levy (*supra* fn. 204), 66.

[247] The direct cost of compensation indeed fell entirely upon the employers. Apart from the prevention argument and the profit motive, this decision apparently resulted from the simple observation that the workers themselves were too poor to be able to provide adequately for themselves and the employers consequentially remained the fittest persons to bear this cost. J.C. Brown, (*supra* fn. 204), 72–73; A. Wilson/H. Levy (*supra* fn. 204), 63.

[248] P.W.J. Bartrip (*supra* fn. 206), 10 and 95; J.C. Brown (*supra* fn. 204), 73.

[249] P.W.J. Bartrip (*supra* fn. 206), 9 and 16; E.P. Hennock (*supra* fn. 218), 79.

[250] P.W.J. Bartrip (*supra* fn. 206), 16.

[251] W.R. Cornish/G. Clark (*supra* fn. 211), 529; A. Wilson/H. Levy (*supra* fn. 204), 66.

English traditions, character and temperament (in contrast with the German).[252]

89   Secondly, the level of compensation that the Act procured, was rather low, since the victim only received 50% of his previous earnings and this up to a certain maximum. Full damages were still something to be sought through Common Law actions, which were not abolished by the Act.[253] On the contrary, the retention of this possibility was explicitly stated in the Act because, according to Chamberlain, there might be 'extremely rare' cases 'in which there was such gross personal negligence on the part of the employer that the compensation awarded by the Bill would be insufficient, and something in the nature of punitive proceedings ought to be contemplated', and because, according to others in parliament, existing legal rights should not be withdrawn not even in return for a generous new one.[254] Another explanation for the low benefits obtainable is of course the heavy opposition of employers who feared the financial consequences of the Act, especially taken into account the no-fault principle.[255]

90   The 1897 Act was anyhow the result of a compromise between various interests groups against the background of a general awareness of the necessity of social legislation taking into account the 'socialist danger': For workmen the Act made compensation a reality; employers accepted it as a necessary evil that they preferred above the possible alternatives; insurance offices saw that new and lucrative business would follow; lawyers liked it because it did not interfere with the sanctity of the common law; and the Conservatives – who were in charge with the largest majority since 1832, saw it as a way of attracting working class support to the detriment of the Liberals.[256]

3.3.2 A Second Shift

91   The 1897 Act was seen as a legislative experiment that had to be evaluated after seven years. The assessment by a Departmental Committee under the chairmanship of Sir Digby in 1904 was cautious optimistic: The act worked reasonably well and had put no crashing burden on industry. As a consequence its recommendations and suggested changes were modest and rather conservative. Except for some streamlining and rendering the law more logical and efficient, they strictly kept to the principals and fundamental methods on which legislation had been hitherto based.[257]

---

[252] E.P. Hennock (*supra* fn. 218), 65 and 74–75.
[253] But the employee could not recover twice; an application for compensation acted as an election not to seek damages: P.W.J. Bartrip (*supra* fn. 206), 219; J.C. Brown (*supra* fn. 204), 74; W.R. Cornish/G. Clark (*supra* fn. 211), 529.
[254] P.W.J. Bartrip (*supra* fn. 206), 219–220.
[255] P.W.J. Bartrip (*supra* fn. 206), 9 and 17; A. Wilson/H. Levy (*supra* fn. 204), 68. In reality, the insurance premiums were indeed higher than expected but not unreasonably or problematically high: P.W.J. Bartrip (*supra* fn. 206), 18.
[256] P.W.J. Bartrip (*supra* fn. 206), 10 and 12; A. Wilson/H. Levy (*supra* fn. 204), 66 and 69.
[257] P.W.J. Bartrip (*supra* fn. 206), 44; W.R. Cornish/G. Clark (*supra* fn. 211), 530; A. Wilson/H. Levy (*supra* fn. 204), 100.

In general, in the first years after its commencement it was acknowledged that 92
the 1897 Act had effected considerable good, but also had many defects in
need of rectification. The complaints were especially directed against the considerable litigation that the Act provoked due to the many uncertainties as to
determining the cases that were entitled to compensation.[258] As the reduction
of litigation, as a means of improving industrial relations, had been one of the
objectives[259] during the preparation of the Act, this was an important objection. Therefore a Bill was proposed to simplify the Act in order to reduce the
uncertainty and decrease the amount of litigation and concomitant expenses.
The 1906 Act[260] would double the scope[261] of the WCA by extending the principle of workmen's compensation to practically all wage-earners. It also introduced compensation for a number of industrial diseases, since judges had
come to accept that a disease might in certain circumstances be regarded as
the direct outcome of an accident.[262]

But the amount of litigation remained high and seemed inherent to the scheme 93
itself.[263] It also became clear that the WCA equally failed at realizing the other
main objectives, compensation and safety. Already the Digby Committee had
expressed doubts with regard to the proof of increased safety due to the Act
and feared that insurances had made the employers more careless.[264] According to the Holman Gregory Committee (1920) the effect of insurance had indeed been the separation of the safety and compensation objectives[265] and it
had to confess that the WCA 'have hitherto included no provision to encourage prevention'.[266] Also the employers denied that the WCA had assisted industrial safety[267] and in the Government's White Paper of 1944 it was concluded that workmen's compensation had been ineffective in advancing either
safety or the rehabilitation of the injured.[268]

As to the compensation goal the Digby Committee of 1904 could still start its 94
evaluation confirming that the Act's aim was not to provide complete indemnity but substantial relief and conclude that no complaints were made[269] as to
the sufficiency of compensation.[270] The only problem seemed to be the risk of
possible insolvency of employers and insurers[271], which the Committee rec-

[258] P.W.J. Bartrip (*supra* fn. 20), 23 and 40.
[259] P.W.J. Bartrip (*supra* fn. 206), 10.
[260] 6 Edw. 7. C58.
[261] The range of application of the WCA was already extended to agricultural labourers by an Act
in 1900.
[262] A. Wilson/H. Levy (*supra* fn. 204), 104. An Order of the Secretary of State would further
extend this list up to 30 diseases: A. Wilson/H. Levy (*supra* fn. 204), 134.
[263] P.W.J. Bartrip (*supra* fn. 206), 60; W. Beveridge, *Social Insurance and Allied Services* (1942), 38.
[264] P.W.J. Bartrip (*supra* fn. 206), 41–42; A. Wilson/H. Levy (*supra* fn. 204), 77.
[265] P.W.J. Bartrip (*supra* fn. 206), 95.
[266] Citation by A. Wilson/H. Levy (*supra* fn. 204), 170.
[267] P.W.J. Bartrip (*supra* fn. 206), 195.
[268] P.W.J. Bartrip (*supra* fn. 206), 201.
[269] Contra: A. Wilson/H. Levy (*supra* fn. 204), 88. See also J.C. Brown (*supra* fn. 204), 17–19.
[270] P.W.J. Bartrip (*supra* fn. 206), 43; A. Wilson/H. Levy (*supra* fn. 204), 84–85.
[271] A. Wilson/H. Levy (*supra* fn. 204), 89–90.

ommended to remedy by regulating private insurers.[272] But among labour organisations compensation and compensation security was gaining importance and further demands were made in that area. The TUC's 1905 Congress showed for the first time a reversal of priorities from safety to compensation. Due to the way in which the insurance companies dealt with claims and claimants[273], the TUC favoured henceforth compulsory State insurance. As to the prevention of accidents one would from now on rely on the Government's control on the carrying out of safety regulations. Another TUC's demand concerning compensation was the abolition of the defence of serious and wilful misconduct. The 1906 Act would indeed limit the application of this rule on the grounds that it should not hinder the worker's capacity to obtain compensation.[274]

95   As already mentioned the Digby Committee was quite cautious in formulating its recommendations; it considered proposing or even discussing the introduction of some system of national insurance 'premature and beyond our commission'. But in its analysis it admitted that the problem of insolvency 'could only be solved by substituting for the personal liability of the individual employer the security of a fund the solvency of which was assured'.[275] And in the same way the Committee indicated that a system of national insurance 'would provide (...) for every employer and every workman *complete security*', that 'ultimately some form of *compulsion* might be adopted', and that 'under such a system *larger benefits* (...) might be provided for'. And it concluded: '(...) beneficial as we believe the legislation of 1897 to have been on the whole, we do not think it can be regarded otherwise than as a step in the direction of a more comprehensive system'.[276] It thus seems that, despite the fact that no official recommendation was made in that sense, the Digby Committee believed that – in particular with regard to the compensation goal – the Workmen's Compensation Act of 1897 represented only a first step in an evolution. It even suggested the instrument by which the ultimate goal could be reached, i.e. a national (compulsory) insurance. The same thing occurred during the parliamentary discussions of the 1906 Act. Many were led to favour compulsory State insurance, not by dogma but out of the recognition that if certain economic and social objectives were to be realised, this was the only means of achieving them.[277] Still, the 1906 Act itself would hold on to the principles of 1897.

96   The Holman Gregory inquiry of 1919 presented a better opportunity for change in that respect. The war experience had strengthened the case for com-

---

[272] P.W.J. Bartrip (*supra* fn. 206), 44.
[273] Insurance offices sent out 'claim settlers' to coerce injured workmen to accept low lump sums as a settlement of the accident. A. Wilson/H. Levy (*supra* fn. 204), 93.
[274] Another argument put forward was the ineffectiveness of the rule in penalizing negligent workmen. P.W.J. Bartrip (*supra* fn. 206), 52.
[275] Cited by A. Wilson/H. Levy (*supra* fn. 204), 91.
[276] Cited (more elaborately) by A. Wilson/H. Levy (*supra* fn. 204), 91. My italics.
[277] P.W.J. Bartrip (*supra* fn. 206), 50–51.

prehensive investigation and had made policy makers and the general public receptive to bolder reforms. As a consequence the Committee's terms of reference were wide and included the investigation of the desirability to establish a system of accident insurance under the control and supervision of the State.[278] The Holman Gregory Committee acknowledged that the fact that a lot of mainly small employers were not insured, endangered the financial certainty of the employee-victim and advocated the compulsory character of insurance, especially since it also proposed increased benefits which should not become illusionary.[279] Taking into account the insurance companies' high profits and expenses (52%) the Committee also favoured state supervision on premium setting so that at least 70% of premium income would be paid out in benefits.[280]

But that was as far as the Committee was willing to go with regard to state intervention. The idea of a State insurance was rejected. The Committee made thereby reference to the trade unions who were opposed to any State scheme when this would involve financial contributions by workers, and to the opposition of employers and insurers who believed that a State monopoly would be more expensive and less efficient than the existing scheme. And in a competition model, the State would attract too high a proportion of 'undesirable risks'. Moreover, the Committee considered that it would be inappropriate for the State to be involved in a system which generated much legal and factual dispute between the different sides of industry.[281] In sum, the Committee appeared, especially taken into account the favourable conditions, rather conservative and failed to establish a new landmark in the history of English Workmen's Compensation.[282]   97

In the subsequent legislation the recommendation for compulsory insurance as well as the proposal for state supervision of insurance premium rates was ignored. However, the Home Office reached a voluntary agreement with the Accident Offices Association that 60% of premium income would be paid in benefits. The introduction of compulsory insurance was considered inopportune taking into account the additional and costly machinery that would be imposed on industry at a time when reconstruction was proceeding.[283]   98

The second shift would eventually take place after the Second World War. The famous Report of the Inter-departmental Committee on Social Insurance and Allied Services published in 1942, the so-called Beveridge Report, would serve as a starting point therefore. Beveridge's aim was to set out the lines of a   99

---

[278] P.W.J. Bartrip (*supra* fn. 206), 88; A. Wilson/H. Levy (*supra* fn. 204), 133–135.
[279] P.W.J. Bartrip (*supra* fn. 206), 91; P. Cohen (*supra* fn. 209), 201; A. Wilson/H. Levy (*supra* fn. 204), 160–163.
[280] P.W.J. Bartrip (*supra* fn. 206), 91; P. Cohen (*supra* fn. 209), 201; A. Wilson/H. Levy (*supra* fn. 204), 164–165.
[281] P.W.J. Bartrip (*supra* fn. 206), 91.
[282] P.W.J. Bartrip (*supra* fn. 206), 92; A. Wilson/H. Levy (*supra* fn. 204), 226.
[283] P.W.J. Bartrip (*supra* fn. 206), 108; P. Cohen (*supra* fn. 209), 202.

Plan of Social Security 'all-embracing in scope of persons and of needs' that would make want under any circumstances unnecessary. The main feature of this Plan was a scheme of social insurance that would cover all citizens and provide non-means tested benefits up to subsistence level against a variety of risks of earnings interruption.[284] Provision for industrial injuries and occupational diseases was to be included within this general framework of social insurance.

100   An interesting element in the Report is the evaluation Beveridge makes of the WCA for this assessment gives an idea of the features Beveridge valued important for a future work-related injuries compensation scheme to display. Beveridge started to stress that the supersedure of the present scheme in his unified Plan for Social Security should not be interpreted as a denial of any good in it. On the contrary, the existing scheme had conferred great benefits in the past and had certain merits: it had provided compensation in the great majority of cases without serious difficulty or delay; it facilitated the return to work by maintaining the link between employer and employee; it had given employers freedom in their arrangements for insuring, so that their liability could be covered on economical lines; and, by allowing premiums to be adjusted to ascertained risks, it had facilitated accident prevention.[285]

101   As Bartrip rightly remarks, it is highly debatable whether these findings accorded to the facts.[286] Anyway, they do reflect the characteristics that Beveridge esteemed essential for any work-related injuries compensation scheme: security of compensation, peaceful industrial relations, rehabilitation, economic efficiency and prevention.

102   Subsequently the Beveridge-Committee also listed the disadvantages of the existing system. This enumeration was largely a confirmation of the defects that had been known and discussed for decades before: the risk of contention and the elevated level of legal and administrative costs; the lack of absolute security of income maintenance due to the absence of compulsory insurance, the unequal position of employer and employee and the practice of settlements and lump sums; and the failure to contribute to the most important purpose of all, the restoration of the injured employee to the greatest possible degree of production and earning as soon as possible.[287] This brought Beveridge to conclude that the system '(…) was based on a wrong principle and has been dominated by a wrong outlook'.[288]

---

[284] W. Beveridge (*supra* fn. 263), 9.
[285] W. Beveridge (*supra* fn. 263), 35–36.
[286] P.W.J. Bartrip (*supra* fn. 206), 180–181.
[287] W. Beveridge (*supra* fn. 263), 36–38. This was indeed one of the weak points of the Workmen's Compensation Act of 1897: it did not include medical or surgical costs and seemed to have the wrong effect on rehabilitation (the greater the recovery, the less the payment): J.C. Brown (*supra* fn. 204), 120–123.
[288] W. Beveridge (*supra* fn. 263), 38.

Instead, Beveridge proposed a State administered compulsory social insurance   103
financed by contributions of employers, employees and the Exchequer, which
would compensate for all physical disability, regardless of cause, with a flat-
rate benefit unrelated to previous earnings. Ideally, according to Beveridge, industrial disability should be treated in the same way as all physical disability,
since 'this would avoid the anomaly of treating equal needs differently and the
administrative and legal difficulties of defining just what injuries were to be
treated as arising out of and in the course of employment'.[289]

However, he also realised that dropping the industrial preference would be un-   104
acceptable to the workmen. Apart from the historical argument (a separate
treatment has been there for over 40 years) Beveridge came up with three reasons to maintain a preferential treatment for work-related injuries lasting longer than 13 weeks. First of all, higher benefits were considered necessary to
make sure that men should enter the dangerous industries that were vital to the
community. Secondly, in contrast with other accidents an industrial injured
had been disabled while working under orders. And thirdly, special provision
for industrial accidents appeared necessary to sustain the case for limiting employers' liability under Common Law.[290] In connection with this, Beveridge also
recommended some variation in financing the scheme. As a departure from the
rule of a general pooling of risks (based on the 'community of interest') a special levy was put on employers in hazardous industries in order to give a definite
financial incentive for prevention of accident and disease.[291] In proposing this,
the Committee actually went back to the thinking behind the 1897 Act.

In general, Beveridge's ideas were not 'notably innovatory', the most funda-   105
mental change being to make compensation for work-related injuries part of
an interrelated system of welfare in which medical treatment and rehabilitation should be primary considerations.[292] The Beveridge-report can indeed be
seen as a coherent and integrated presentation of ideas and suggestions for a
more comprehensive reform of workmen's compensation that had been existing for years but never had been realised nor fully endorsed, such as workers'
contribution[293], compulsory insurance[294] and state participation. That they

---

[289] W. Beveridge (*supra* fn. 263), 38–39.
[290] W. Beveridge (*supra* fn. 263), 39. For a critical assessment of these arguments, see: P.W.J.
Bartrip (*supra* fn. 206), 184; N.J. Wikeley/A.I. Ogus/Barendt, *The Law of Social Security* (5th
edn. 2002), 717–718.
[291] W. Beveridge (*supra* fn. 263), 41–42.
[292] P.W.J. Bartrip (*supra* fn. 206), 187.
[293] The financial contribution of workers was not only motivated as a direct consequence of the
reversion of the scheme into a social insurance, but was also presented as a matter of balances.
Employees would in return obtain an equal weight in the administration of the scheme, and for
employers it was meant to reciprocate for the loss of control over administration and their
right to go to the Courts for settlement of disputes: P.W.J. Bartrip (*supra* fn. 206), 207.
[294] As to compulsory insurance an important precedent had been made by the approval of Nichol-
son's Bill that introduced compulsory insurance in the coal industry, the *Workmen's Compensation (Coal Mines) Act 1934*. See: P.W.J. Bartrip (*supra* fn. 206), 157–162; A. Wilson/H.
Levy (*supra* fn. 204), 257–267.

could be realized now and all in one time was mainly due to the war experience that had brought about a new sense of social solidarity and produced a unique consensus on the theme in parliament.[295]

106   Beveridge's criticisms of the WCA as well as his proposal of unification within a global system of Social Security, were generally endorsed by subsequent committees and the government.[296] However, the special levy on employers[297] in hazardous industries was rejected. Although the safety incentives of the measure were acknowledged in theory, it was thought to have little effect in practice owing to the effect of insurance and relatively low cost of accidents to employers.[298] It was also considered an inappropriate principle in a Social Security scheme, a perpetuation of the industrial antagonism of workmen's compensation and a duplication of administrative machinery. Moreover, employers feared that it would encourage unions continually to demand higher benefits.[299]

107   The retention of the industrial preference on the other hand did receive full support (under pressure of the TUC)[300] – officially on the grounds that in 'the industries where most industrial accidents occur, workmen are exposed to far greater risks than citizens in the ordinary walks of life'[301] – although some members of parliament raised the question of conformity with the principle of equal treatment and the practical problem of eligibility.[302]

108   With these adjustments the Beveridge-report would ultimately result in the introduction of the *National Insurance (Industrial injuries) Act* 1946[303] (implemented on 5 July 1948), which would due to the electoral success of Labour[304] and to trade unions pressure, provide yet more generous benefits than originally foreseen. A newly set up *National Health Service* (Acts of 1946–47) would offer a comprehensive scheme of medical, surgical and other health care.

109   To one important question this new Industrial Injuries Act gave no answer: the possibility of alternative remedies. In his Report Beveridge recommended to have this question dealt with in a particular committee.[305] Since starting from the 1930s more claims were being made under Common Law due to changing

---

[295] J.C. Brown (*supra* fn. 204), 21 and 24.
[296] P.W.J. Bartrip (*supra* fn. 206), 194, 198 and 201; J.C. Brown (*supra* fn. 204), 27.
[297] The TUC wanted the whole cost of compensation entirely upon the employers. These latter were divided on the matter. However, it was clear to them that if they were to be the only party to finance the system, they were also to have a monopoly on its administration: J.C. Brown (*supra* fn. 204), 75.
[298] J.C. Brown (*supra* fn. 204), 109.
[299] P.W.J. Bartrip (*supra* fn. 206), 196–197.
[300] What's more, the limitation to cases lasting more than 13 weeks was dropped.
[301] P.W.J. Bartrip (*supra* fn. 206), 196 and 202.
[302] P.W.J. Bartrip (*supra* fn. 206), 205–206.
[303] 9 and 10, Geo. VI, C62.
[304] In 1945 a Labour government was elected.
[305] W. Beveridge (*supra* fn. 263), 103.

judicial attitudes and important legislative changes[306], this was indeed a matter of high importance. Beveridge himself did not want to make definite choices but made clear that leaving the employer's common law liability unchanged, was irreconcilable with the provision of compensation by way of social insurance and very different of the practice of other countries.[307] Other arguments were the fact that employers could not be expected to pay twice, the desirability to remove the compensation issue from the contentious atmosphere of the Courts, and the belief that the realities of modern industrial production made it impossible for the employer to control and supervise employees and thus unfair for him to be held morally and legally responsible for an employee's negligence. Arguments in favour of retention were the possibility of full compensation, the influence in improving safety and the equal position with other members of the public with regard to the possibility of legal action.[308]

In its final report of 1946 the Monckton Committee recommended to retain 110 the right of action for damages for personal injury but with the limitation that national insurance benefits would be deducted from common law damages. The argument put forward for this choice was however not the familiar one of promoting safety, but the desire to avoid 'under-compensation' of those who might otherwise have had a claim against a negligent employer. With regard to the employer's liability for breach of statutory duty, the Committee admitted a defence of having taken all reasonable care.[309]

Eventually, the *Law Reform (Personal Injuries) Act* that would be enacted on 111 30 June 1948 would not follow these recommendations but was much more in line with the TUC[310] views: A cumulation of damages under Common Law and national insurance benefits would be possible during five years up to one half of the value of industrial injury benefit.[311] According to Cornish & Clark this extra provision could be seen as the reflection of the employee's personal contribution to the insurance fund.[312] In general, it was clear that the Act was the result of a compromise.[313]

Although the shift towards a social security regime cannot be denied, the in- 112 troduction of the *Industrial Injuries Scheme* (IIS) of 1948 did thus not imply

---

[306] Due to the fact that breach of statutory duty could be presumed to be negligence, the introduction of the *Factory Act of 1937* for example – which imposed stricter safety obligations upon the employer – widened considerably the changes for successful litigation for the employees. P.W.J. Bartrip (*supra* fn. 206), 221–223.
[307] W. Beveridge (*supra* fn. 263), 46.
[308] P.W.J. Bartrip (*supra* fn. 206), 226–228; W. Beveridge (*supra* fn. 263), 46 and 101–102.
[309] P.W.J. Bartrip (*supra* fn. 206), 227; W.R. Cornish/G. Clark (*supra* fn. 211), 537–538.
[310] The employers for their part were in favour of a restriction of the possibility of a tort action to exceptional cases. They considered this a reasonable offset against their substantial financial contribution to the system and as beneficial to industrial relations. J.C. Brown (*supra* fn. 204), 159.
[311] P.W.J. Bartrip (*supra* fn. 206), 229–230.
[312] W.R. Cornish/G. Clark (*supra* fn. 211), 538.
[313] N.J. Wikeley, *Compensation for Industrial Disease* (1993), 66–67.

giving up on civil liability law as a way of compensating for industrial injuries. On the contrary, it would constitute more than ever a valuable alternative remedy for victims-employees since their legal position was much improved by post-war legislation. Former obstacles were removed[314] and duties of care[315] were imposed on the employer, providing the employees a possible action for breach of statutory duty.

113   These changes were all motivated by the argument that they would promote safety.[316] It is however questionable whether these measures really have contributed a great deal to improved safety, especially after 1969 when liability insurance, the possibility of which theoretically blocks the deterrent effect of tort law[317], was made compulsory by the *Employers' Liability (Compulsory Insurance) Act*. This Act introduced a system of automatic compensation but is still based on the civil liability of the employer, since claims depend on the establishment of negligence of the employer. The combined effect of all these changes was that, following the introduction of the IIS, Common Law actions became much more common (and successful) instead of falling into disuse.[318]

3.3.3 Recent Developments

114   The English compensation scheme as it was developed in 1946 remained largely unaffected during the following decades. In the 1970s an administrative change took place that more strongly integrated the IIS into the general structure of social security. Under the 1973 and 1975 Social Security Acts, the Industrial Injuries Fund was merged with the National Insurance Fund and National Insurance (Reserve) Fund to form a single unit. Administrative simplification, a general dislike of earmarked contributions and especially the healthy surplus of the Industrial Injuries Fund seem to have been the main motive for this action.[319] The TUC however looked upon this change with sorrow as it would render a future abolition of the employees' contribution or the introduction of premium differentiation towards companies more difficult.[320]

115   In 1978 the Pearson Committee that had been asked to make a global assessment of the compensation system in place, issued its report. In general, the judgement on the IIS was positive. The Committee found that most, including the TUC, approved of the system and stated that there were few countries that could in all re-

---

[314] The defence of common employment was formally abolished by *The Law Reform (Personal Injuries) Act of 1948*, and that of contributory negligence severely restricted by *The Law Reform (Contributory Negligence) Act of 1945*. *The Legal Aid and Advice Act of 1949* that offered legal assistance to those with a low income, took down another important barrier.

[315] The culmination of which would be the *Health and Safety at Work Act of 1974* imposing upon the employer a general responsibility to ensure, so far as is reasonably practicable, the health, safety and welfare at work of all his employees.

[316] J.C. Brown (*supra* fn. 204), 99.

[317] Cf. *supra* no. 8.

[318] J.C. Brown (*supra* fn. 204), 37, 99 and 161; B.S. Markesinis/S. Deakin/A. Johnston, *Markesinis and Deakin's Tort Law* (2003), 561.

[319] J.C. Brown (*supra* fn. 204), 86; N.J. Wikeley/A.I. Ogus/Barendt (*supra* fn. 290), 715.

[320] J.C. Brown (*supra* fn. 204), 87.

spects match provision in the UK.[321] Still, the system could benefit from some harmonisation and simplification, and should be improved and extended to include for instance commuting accidents and the self-employed.[322] The Committee also recognized that differential contributions as proposed by the TUC (but opposed by the employers)[323], could be a strong incentive to establish and maintain safe working conditions, but considered it too difficult to realize in practice taking into account the high costs and administration in relation to the probably marginal effects.[324] The Pearson Committee admitted that Beveridge's arguments to endorse the industrial preference weighted a good deal less at the time, but nevertheless wanted to preserve the existent level of benefits.[325]

A great deal of attention was paid by the report to the relationship between compensation via tort and state provision. After a quite elaborate assessment of the advantages and disadvantages of each of the two compensation systems, the Committee reaches the conclusion that both should supplement each other in a mixed system of compensation. In the words of the Committee: 116

'We hope that, after this report, it should no longer be possible to think that there is no relation between compensation for the injured and bereaved provided by the Social Security system on the one hand and that provided by the Common Law of tort on the other. The two systems work in different ways and operate from different philosophies, and tort will become the junior partner. But we have shown how, by fitting them together more effectively, compensation for injured people and their dependants can be appreciably improved and extended'.[326]

According to the Committee, tort should thus be retained as a supplementary means of compensation. Its abolition would only be justified if the no-fault scheme could be significantly improved, which was not the case.[327] Other arguments in favour of preservation are the deterrent effect[328] of tort (through surveys by insurance companies and due to the exposure on trial) and the idea that workmen should not be denied a normal civil right.[329] But the Committee is equally clear regarding the exact position of tort within this mixed compensation system. It proposes 'a considerable shift of emphasis from the tort to 117

---

[321] Pearson Report, *Report of the Royal Commission on Civil Liability and Compensation for Personal Injury, vol. 1* (1978), 173.

[322] The self employed were considered as deserving as the workmen. The extension to accidents to and from work only got a just majority but was demanded by the TUC and ILO recommendation 121. Pearson Report (*supra* fn. 321), 184–186.

[323] J.C. Brown (*supra* fn. 204), 111.

[324] Pearson Report (*supra* fn. 321), 91–192.

[325] Pearson Report (*supra* fn. 321), 70.

[326] Pearson Report (*supra* fn. 321), 364.

[327] Pearson Report (*supra* fn. 321), 194.

[328] The Committee was however not blind for the results in the Safety and Health at work-report (1972) of the Robens Committee. This Committee found that compensation litigation could also have a counterproductive effect on industrial safety, that is when the implications of safety measures are evaluated not in terms of prevention and safety but in terms of limiting compensation costs.

[329] Pearson Report (*supra* fn. 321), 193–194.

the social security system of compensation' and recommends that 'Social Security should be recognized as the principal means of compensation'.[330] Double compensation should therefore be avoided by offsetting social security benefits in the assessment of tort damages.[331] The Committee finally also asked to
extend the range of those receiving compensation and to spend the available
money on the more serious injuries rather than on the minor injuries.[332]

118   The government accepted this view in principle in her 1981 White Paper. Such
a scheme would indeed have the advantage of not lessening the employers' liability and would recover sums that might finance improved provision for all
injured workers. But since the government also feared a considerable increase
in staff numbers and practical difficulties in the vast majority of cases which
were settled out of court, it nevertheless concluded that for the time being direct recovery was not an option which could usefully be pursued.[333]

119   Reports that demonstrated which sums could be recovered when deducting
national insurance benefits from common law damages increased the pressure
and finally brought the government to introduce a recoupment scheme in the
Social Security Act of 1989. All social security benefits received in the five
years after the injury or the date of settlement (whichever came first) should
be deducted from all tort damages (including damages for pain and suffering
as well as loss of earnings) but paid over to the Secretary of State.[334] The reasoning was that the state should not subsidise defendants and that accident
victims should not get this windfall from double compensation payments.

120   The scheme was however strongly opposed by groups as the TUC, the Con-
federation of British Industry and the Association of British Insurers. In 1988
also the Industrial Injuries Advisory Council (IIAC)[335] had defended its opposition to the full recovery of benefits. Since employees and employers both
contributed to the National Insurance Fund it was considered unfair to take
benefits entirely into account. Moreover, the Council argued (referring to literature) that the assumption that a tort award would fully meet the accident victim's financial needs in respect of past, present and future losses could not be
sustained.[336] In 1997 the recovery scheme was amended by the Social Security
(Recovery of Benefits) Act. To make the scheme more fair compensation paid
in respect of pain and suffering was from now on to be discounted.[337]

---

[330] Pearson Report (*supra* fn. 321), 363 and 367.
[331] Pearson Report (*supra* fn. 321), 68 and 367.
[332] Pearson Report (*supra* fn. 321), 367.
[333] N.J. Wikeley (*supra* fn. 313), 67–68.
[334] In addition, there was no longer any discretion as to whether or not the benefits would be
deducted and neither were the provisions confined to court awards.
[335] The statutory provision for the Council is currently contained in sec. 171 of the Social Security Administration Act 1992. Its task is to provide independent advice to the Secretary of
State on matters concerning the industrial injuries scheme.
[336] N.J. Wikeley (*supra* fn. 313), 68–69.
[337] S. Jones, Social Security and Industrial Injury, in: N. Harris, *Social Security Law in Context*
(2000), 490. This suggestion had already been made more than ten years before: see N.J.
Wikeley (*supra* fn. 313), 70.

In her 1981 White Paper the government also envisaged a restructuring of the   121
system so that it would focus on the more seriously disabled.[338] During the
1980s and early 1990s a series of changes would be carried through that
would result in a strongly simplified system with a disablement benefit offering compensation for 'loss of physical or mental faculty' and two extra sup-
plements for cases of the utmost severity.[339] All the other existing benefits
were abolished. According to the government these changes represented 'a
sensible further step towards a more coherent system of benefits for sick and
disabled people'; in reality however they could be seen as economy measures
that encountered strong opposition from the trade unions and the IIAC.[340]

Another important element in the recent discussion on compensation for in-   122
dustrial injuries is whether or not to keep the 'industrial preference'. This
question followed among others from the Department of Health and Social
Security (DHSS)-discussion document of 1980, in which the high administrative cost of the separate system was denounced given the fact that the benefits
were hardly more generous than the general social security benefits.[341]

In 1990 the IIAC judged it necessary to issue a position paper on this topic.   123
After recalling the generally expressed main disadvantages of upholding a
separate scheme[342], the IIAC strongly argued in favour of retention.[343] As a
first argument for this stand, the Council refers to the fact that also in the
present-day society risk and unsafe working conditions cannot be eliminated
altogether[344] and that as a result society has a continued duty to provide for the
consequences of industrial injury.[345]

The IIAC further believes that neither Common Law arrangements nor the   124
general social security provision can meet in full the demands that according
to the Council should be made to a scheme for industrial injury.[346] Even if the

---

[338] N.J. Wikeley (*supra* fn. 313), 62; N.J. Wikeley/A.I. Ogus/Barendt (*supra* fn. 290), 716.

[339] *The Constant Attendance Allowance and the Exceptionally Severe Disablement Allowance* (SSCBA 1992, sec. 104 and sec. 105).

[340] J.C. Brown (*supra* fn. 204), 91; N.J. Wikeley (*supra* fn. 313), 64; N.J. Wikeley/A.I. Ogus/ Barendt (*supra* fn. 290), 716.

[341] S. Jones (*supra* fn. 337), 465; N.J. Wikeley (*supra* fn. 313), 62; N.J. Wikeley/A.I. Ogus/ Barendt (*supra* fn. 290), 716.

[342] No justification for differentiating between disabled people, too complex, significant extra costs and inadequate level of the provisions; Industrial Injuries Advisory Council (IIAC), *The Indus-trial Injuries Scheme and the Reform of Disability Income. Position paper No. 5* (1990), 4–5.

[343] This plea does not prevent the IIAC of criticizing the system in place. The Council recognises 'that there is a case for substantial rationalisation, and a need to upgrade a number of the existing income benefits and to recognise more fully the costs of disability' (IIAC (*supra* fn. 342), 5).

[344] The Council points to some features of our contemporary economy that create new areas of risk: the development of the 'flexible' labour market with its greater use of sub-contracting and self employment, the emergence of new small firms inexperienced in safety matters, and more generally the economic pressure to compete (IIAC (*supra* fn. 342), 7).

[345] Making this argument the Council mainly falls back on the reasons given by Beveridge: the essential character of some high risk jobs to society or to the economy, and the limited control of the employee over his/her work environment (IIAC (*supra* fn. 342), 6). Cf. *supra* no. 104.

[346] IIAC (*supra* fn. 342), 8–10.

general provision for sickness and disability was reformed, it could not be expected that a general scheme would provide compensation for the actual injury, or that it would include a benefit for reduced earnings arising from the injury, of the type needed in an occupational scheme. According to the Council 'this was not an argument for an industrial preference of the type once prevailing', but an argument 'against the assumption that, because a scheme of similar standards is unlikely to be offered to all sick or disabled people, it should not be offered to those with occupational injuries'.[347]

125 A third important argument put forward by the IIAC in favour of retaining a separate scheme is its contribution to prevention. This contribution is made through the provision of a benefit for reduced earnings[348], the identification of the physical and financial consequences of occupational injury, and through the work of the Council itself in identifying occupationally caused disease.[349] Since the cost of the scheme is already being met through the National Insurance contributions of both sides of industry, this need not be seen as an obstacle. Nevertheless it is the Council's opinion that, given that society has a direct interest that work be undertaken, there would be good reasons for the subvention from the general taxpayer to be restored.[350]

126 The Council finally adds the European dimension as an argument. Starting from the observation that 10 out of the 12 Member States have a separate scheme, it argues that the retention of such a scheme in Britain would serve the goals of harmonisation of legislation and the protection for mobile workers.[351]

127 In reality it is of course mainly a political fact that it is hard to remove privileges once assigned to groups with a significant voting power.[352] Still, some government measures have resulted in a small decrease of the industrial preference. First of all this counts for the benefits and allowances of the IIS that were abolished. Important was the abolition of the industrial injury benefit in 1982. Accident victims had now to fall back on the ordinary sickness benefit or statutory sick pay (SSP) during the first 15 weeks of their disablement. SSP[353] was moreover partially privatized when the possibility for employers to get reimbursed from the National Insurance Fund was reduced to 80%.[354]

---

[347] IIAC (*supra* fn. 342), 10.

[348] At the time the report was published (January 1990), the *Reduced Earnings Allowance* (REA) that had come to replace the *Special Hardship Allowance* in 1986, and paid extra compensation to injured employees who had returned to work but who were unable to regain their previous level of earnings, had not yet been abolished. This was only to be the case for new claims after October 1990 (N.J. Wikeley (*supra* fn. 313), 64).

[349] IIAC (*supra* fn. 342), 10–11.

[350] IIAC (*supra* fn. 342), 12.

[351] IIAC (*supra* fn. 342), 13.

[352] S. Jones (*supra* fn. 337), 494; N.J. Wikeley/A.I. Ogus/Barendt (*supra* fn. 290), 718.

[353] This is the obligation for the employer, introduced in 1983, to pay sick pay for the first 28 (originally 8) weeks of incapacity.

[354] This reduction did not count for small employers: N.J. Wikeley (*supra* fn. 313), 59.

From 1986 on no more benefits were given for injuries that counted for less than 14%[355] and since 1997 there is a time-limit of three months for claiming the disablement benefit. Although the possibility to repeal entitlement to the remaining additional allowances is already provided for[356], there is no evidence of an immediate government intention to abolish or modify what is left from the industrial preference, except to transfer the cost of payment from the state to the employer and the employer's insurance company.[357]

Since there is wide agreement that the role of the IIS in rehabilitation has been fairly marginal[358], one has also started in the late 1990s to look for ways to improve the connection between compensation, rehabilitation and prevention of industrial injuries, but no immediate plans for reform have so far resulted from this.[359]

128

In the most recent years also the compulsory Employers' Liability Insurance came up as an element for possible reform within the industrial injuries compensation scheme, due to the significant price increases[360] from the year 2000 on. According to the government the problem was not so much the actual level of the premiums – which were said to be reflecting the true economic costs of industrial injuries and still being much lower than those of the international competitors – but the speed of the market adjustment[361] combined with a lack of forewarning.[362] One could not conclude then that the Employers' Liability insurance market had failed but simply that it had not been working well enough. Remarkably, since the compulsory insurance was introduced as a means to increase safety,[363] is that one of the market defects is said to be the fact that premiums often not reflect individual health and safety efforts.[364] Interesting (but maybe not surprising[365]) is also that the rising (legal) costs of resolving claims is indicated as one of the causes of the market adjustment.[366]

129

---

[355] Except in the case of three lung diseases: N.J. Wikeley (*supra* fn. 313), 62. The possibility for entitlements to a lump sum or gratuity for assessments of between 1–19% was introduced in 1953 and it was estimated that in 1983 90% of all assessments came within this lower range (S. Jones (*supra* fn. 337), 481).

[356] SSCBA 1992, sec. 104 (3).

[357] S. Jones (*supra* fn. 337), 493.

[358] J.C. Brown (*supra* fn. 204), 130.

[359] N.J. Wikeley/A.I. Ogus/Barendt (*supra* fn. 290), 717. The government has also reaffirmed this commitment to make rehabilitation play a more central role in the worker's compensation system in its recent reports on the Employers' Liability Compulsory Insurance. Department for Work and Pensions, *Review of Employers' Liability Compulsory Insurance: First Stage Report* (2003), 5.

[360] The average increase was estimated around 40%, but could run up to even 160% for smaller firms and in some sectors: cf. Department for Work and Pensions (*supra* fn. 359), 6.

[361] Before, premiums would have been held 'artificially low'.

[362] Department for Work and Pensions, *Review of Employers' Liability Compulsory Insurance: Second Stage Report* (2004), 4.

[363] Cf. *supra* no. 113.

[364] Department for Work and Pensions (*supra* fn. 362), 4.

[365] Since the Employer' Liability Compulsory Insurance is tort-based (in contrast with compulsory insurances for industrial injuries in countries like Belgium and Germany) and tort law is known as a contentious system (cf. *supra* no. 10).

[366] Department for Work and Pensions (*supra* fn. 359), 8.

130   The existence of the compulsory insurance itself was not under discussion.
       The government stressed that the insurance 'had served (...) well for nearly
       thirty years' and repeated the arguments in favour of the insurance: It offered a
       secure and fair compensation to injured employees, provided an incentive to
       reduce accidents at work and offered businesses the possibility of spreading
       the risk of industrial injuries.[367]

3.4   The Shifts in Belgium

3.4.1   A First Shift?

131   A general awareness of the precarious situation of workmen had been present
       in Belgium since halfway the 19th century. But as the newly arisen state was
       still anxious to guard its distinct liberal fundaments, it showed more reserve
       than other industrialized countries with regard to state interference with work-
       ing conditions. The social riots of 1886 in Liège, Charleroi and the Borinage
       however brought a change in this attitude.[368] The same year a congress was or-
       ganized in Liège and a governmental *Commissie van de arbeid* (Labour com-
       mittee) was installed both with regard to industrial accidents. Four years later
       the Belgian parliament would comply with the King's request to organize a re-
       lief fund for victims of industrial accidents[369] and take into consideration a
       first legislative proposal concerning work-related injuries.[370]

132   Wide agreement existed about the problem to be tackled. The 'Roman' con-
       ception of civil liability as it was codified in article 1382 of the *Burgerlijk Wet-
       boek* (Belgian Civil Code, B.W.) and based on the existence of an agricultural
       society, was not fit to serve as a framework to organize actual industrial rela-
       tions and left the victims of industrial injuries too often without compensation.
       The same result followed from contractual liability law since the labour con-
       tract was hardly regulated.[371] Civil liability left the consequences of accidents

---

[367] Department for Work and Pensions (*supra* fn. 362), 5.
[368] U. Deprez *et al.*, Van steun-en voorzorgskas tot fonds voor arbeidsongevallen, [1990] *Belgisch
       Tijdschrift voor Sociale Zekerheid* (BTSZ) 23, 384; J.-F. Leclercq, Rapport Introductif, in: J.L.
       Fagnart (ed.), 1903–2003. *Accidents du travail: cent ans d'indemnisation* (2003), 1–2; P. Van Der
       Vorst, Esquisse d'une théorie générale du 'risque professionnel' et du 'risque juridique', [1975]
       Journal des Tribunaux (JT), 373; V. Vervliet, *De buitencontractuele aansprakelijkheid bij profes-
       sionele risico's: een onderzoek naar de betekenis en de draagwijdte van de buitencontractuele
       aansprakelijkheid doorheen de evolutie van de professionele risicoverzekering* (2006), 24–25.
[369] The Act of 21 July 1890 was adopted to the occasion of the 25th anniversary of the accession
       to the throne of King Leopold II. The fund it set up, the *Steun- en Voorzorgskas* (Financial Aid
       and Prevention Fund), had to encourage the use of industrial injuries insurances and give relief
       to victims of industrial accidents. G. Delvaux *et al.*, *Honderd jaar sociaal recht in België*
       (1987), 105; U. Deprez *et al.*, [1990] BTSZ 23, 371 and 385.
[370] *Hand.* Kamer 1889–1890, 17 May 1890, 1469.
[371] M. Demeur, Le Risque Professionnel. Traité théorique et pratique de la Loi du 24 décembre
       1903 sur la Réparation des Dommages résultant des Accidents du Travail, [1908] *Pandectes
       Belges* (Pand. Belges), 137; J.L. Fagnart, Un régime nécessaire et exemplaire de réparation
       des atteintes à l'integrité physique. Rapport de synthèse, in: J.L. Fagnart (ed.), 1903–2003.
       Accidents du travail: cent ans d'indemnisation (2003), 307; H. Lenaerts, *Inleiding tot het*

caused by coincidence or *force majeure* or with an unknown cause to the vic-tim-employee. 'Naturally' the employee also bore the damage that resulted from his own (even minor) fault. Since the industrial setting had made em-ployees accustomed to the conditions of danger and to a certain degree also careless – and this to the advantage of the employer, also this latter legal conclu-sion was considered as too far removed from reality to be just. The same remark could be made with regard to the employer's duty of surveillance. Compensa-tion could only be obtained when a fault of the employer could be proven. This burden of evidence was extremely heavy for workmen without any legal knowl-edge or awareness. Workmen were often afraid to press charges against their employer or to testify to his detriment. Moreover, the procedure was slow and jurisprudence instable, indecisive and even contradictory.[372] And at the end of a successful procedure, the employee always ran the risk of being confronted with the insolvency of his employer. First party insurance was no option as it was too expensive taking into account the revenues of workmen.[373]

Lawyers first tried to find a solution to these defects within the existing system of liability law.[374] Sainctelette argued that the employer's obligation to guaran-tee the security of his employees was a contractual one and arose from the tac-it intentions of the contract parties. He saw this obligation as a corollary to the authority the employer disposed of both with relation to the working condi-tions as to the person of the workman himself.[375] He went on interpreting this obligation as a commitment of result, meaning that the occurrence of the acci-dent sufficed to entail the employer's shortcoming. Sainctelette's theory thus resulted in a reversion of the burden of proof. It became the employer who had to prove that he was not responsible for the accident damages. The solution that Sainctelette[376] offered, was without any doubt to the advantage of work-men[377]; however, its legal basis appeared somewhat artificial and was not fully accepted by the *Hof van Cassatie* (Belgian Supreme Court, Hof)[378] that ex-pected a solution from the legislator.

133

---

sociaal recht (1995), 146; Y. Mondelaers, *Het onstaan en de evolutie van de arbeidsongeval-lenverzekering in België (1880–1930)* (2000), 17–21; F. Tanghe, Het arbeidsongeval: aanspra-kelijkheid als sociale constructie, [1989] *Recht der Werkelijkheid* (RdW) 2, 61; P. Van der Vorst, [1975] JT, 374; *Parl. St.* Kamer 1900–1901, no. 123, 1.

[372] Jurisprudence had however become more forthcoming towards the victims by means of a larger interpretation of the employer's duty of care towards the employees: *Parl. St.* Kamer 1901–1902, no. 302, 4; A. Beeckman, Responsabilité des maitres et patrons. Les assurances contre les accidents du travail au Congrès de Liège, [1886] *Journal du Tribunaux* (JT) 367, 1170; Y. Mondelaers (*supra* fn. 371), 24; V. Vervliet (*supra* fn. 368), 40–46.

[373] A. Beeckman [1886] JT 367, 1170–1171; M. Demeur [1908] Pand. Belges, 137–139; J.L. Fagnart (*supra* fn. 371) 309–310; Y. Mondelaers (*supra* fn. 371), 22; P. Van der Vorst [1975] JT, 374–375.

[374] J.L. Fagnart (*supra* fn. 371), 310–311; P. Van der Vorst [1975] JT, 376–377.

[375] Ch. Sainctelette, *De la responsabilité et de la garantie (accidents de transport et de travail)* (1884), 117 and 153.

[376] In France, the same theory was defended by Sauzet.

[377] Accidents with an unknown cause would now have to be carried by the employers.

[378] The theory was for the first time rejected in the judgement of 8 January 1886 (*Pas.*, I, 38) and even more clearly in the judgement of 28 March 1889 (*Pas.*, I, 161).

134   Another theory that attempted to offer workmen an easier possibility for com-
        pensation, was the stand that article 1384 B.W., which constituted the basis for
        some situations of vicarious liability, contained a general principle of faultless
        liability for damage causing objects.[379] This theory was in a first phase also re-
        jected[380] by the Hof but ultimately accepted[381] in 1904, six months after the
        first Belgian *Arbeidsongevallenwet* (Industrial Injuries Act, AOW 1903) was
        promulgated.

135   Although both proposals clearly disposed of some advantages for the victims
        of industrial accidents, they left some categories of accidents uncompensat-
        ed,[382] sat uneasy with the liberty of contract and would not be able to prevent
        long and uncertain legal procedures to arise.[383] As a consequence it soon be-
        came clear to some that what was needed was not a reform of civil law but a
        reform of social law.[384] The basis for such a reform was found in the theory of
        the *risque professionnel*, the idea that industrial accidents inevitably make part
        of the normal risk inherently connected with running a business and therefore
        must be taken into account by the employer as a cost of production.[385] As indi-
        vidual behaviour seemed to have no definite nor measurable impact on the
        global number of industrial injuries, the quest for individual responsibility ap-
        peared to a great extent senseless. Accidents and injuries were a social evil
        closely connected to industrial production and not reducible to the employer's
        free will. Therefore the needs of the victims deserved priority. Their right to
        compensation did not follow from the reprehensible behaviour of the employ-
        er but from the wish to establish a balance between the interests of employers
        and victims. The *risque professionnel* allowed to solidarize producers and
        consumers by turning the premium of industrial insurance into a production
        cost.[386]

136   The proponents of the *risque professionnel* theory indeed considered compul-
        sory industrial insurance, paid for by the employers, as the solution to the
        problem of compensating work-related injuries. Compensation would then be
        paid when the existence of damage linked to the occurrence of an industrial
        accident was demonstrated, irrespective of the question who was responsible

---

[379] J.L. Fagnart (*supra* fn. 371), 311; P. Van der Vorst [1975] JT, 376–377.
[380] The judgment of 28 March 1889 (*Pas.*, I, 161).
[381] The judgment of 26 May 1901 (*Pas.*, I, 246).
[382] Those that were caused by coincidence, *force majeure* or by the fault of the (co-)employee himself. This was also the majority of the cases.
[383] A. Beeckman [1886] JT 367, 1171–1172; Y. Mondelaers (*supra* fn. 371), 37–38.
[384] A. Beeckman [1886] JT 367, 1173; *Parl. St.* Kamer 1901–1902, no. 302, I. Whether the AOW 1903, which would eventually be the result of this reform, was part either of private (civil) law or of public (social) law, would remain a sensitive topic of discussion. As a reaction to the reproach that the new Act had a 'socialist' or 'collectivist' character and as an attempt to broaden its public support, one tried to prove that the new arrangement had to be situated within private (civil) law. The fact that the Act did not impose a tax and that the state did not pay out the benefits itself, were important elements in the argumentation. *Parl. St.* Kamer 1901–1902, no. 302, 6–9; H. Lenaerts (*supra* fn. 371), 146.
[385] U. Deprez *et al.* [1990] BTSZ 23, 385; H. Lenaerts (*supra* fn. 371), 144–145.
[386] M. Demeur [1908] Pand. Belges, 140; F. Tanghe [1989] RdW 2, 62–63.

for the accident to happen. This so-called 'German solution' would give assurance to the victims with regard to compensation, to the employers with regard to the damages to be paid, and to society regarding industrial conflict.[387] It is remarkable that all legislative proposals[388] took insurance (though not always compulsory) as a starting point. The introduction of compulsory insurance based on the *risque professionnel* idea and following the German model had also been the solution already suggested by the *Commissie van de Arbeid* in 1886.[389]

Not everyone however was convinced of this new social law. Some lawyers and parliamentarians wished to hold on to individual responsibility as the essence of morality and dignity of the subject and kept adhering to traditional liability law. They argued that the proposed reform would deprive both employer and employee of their freedom and dignity, as the employer's moral duty to compensate and the victim's right to obtain what he is entitled to, would be replaced by the product of 'taxation'.[390] Moreover, the 'automatic character' of this new compensation scheme would kill providence, provoke indifference and increase the number of accidents.[391]

137

The reformers won the battle and in 1903 the first Belgian AOW was adopted.[392] The events of 1886 that had triggered off the first initiatives with regard to work-related injuries already made clear that the main aim of the Act would be appeasement. To this effect, the realisation of two objectives were considered crucial: securing compensation to the victims in all possible cases and preventing as much as possible legal procedures to arise.[393] The explanatory memorandum confirmed that the victim's compensation should not be dependent upon the employer's precaution or goodwill, but that equity and humanity urged it to be a right for the employee. And it immediately added that the proposed arrangement was also desired by employers in order to avoid irritating procedures[394] and unexpected and uncertain financial costs.[395] It is indeed true that employers were as interested as the employees in replacing the compensation scheme due to the increasingly stringent jurisprudence, the concomitant risk of having to pay full compensation and the contentious character of civil litigation, although the formation of *bedrijfskassen* (company funds) and the

138

---

[387] A. Beeckman [1886] JT 367, 1174.
[388] For a complete overview: M. Demeur [1908] Pand. Belges, 127–136.
[389] U. Deprez *et al.* [1990] BTSZ 23, 384.
[390] H. Lenaerts (*supra* fn. 371), 146–147; Mestreit, Discours de rentrée, [1905] *Jurisprudence Liège* (Jur. Liège), 297 *et seq.*; F. Tanghe [1989] RdW 2, 64.
[391] F. Tanghe [1989] RdW 2, 64.
[392] Date of commencement: 1 July 1905.
[393] M. Demeur [1908] Pand. Belges, 141; F. Tanghe [1989] RdW 2, 67.
[394] Victims-employees had moreover better changes to a successful legal procedure since from 1878 on they had the possibility to start a procedure in civil law to the occasion of a criminal charge, which largely mitigated the burden of proof. See J. Van Langendonck, Arbeidsongeval en aansprakelijkheid, [1988] *Tijdschrift voor Sociaal Recht* (T.S.R.), 75–76 and V. Vervliet (*supra* fn. 368), 32.
[395] *Parl. St.* Kamer 1900–1901, no. 123, 2.

possibility of taking out private insurance had proven to be a valuable alternative to some of them.[396]

139   In fact, the AOW awarded the victim of each injury that occurred during and
      due[397] to the execution of his labour contract a right to compensation. Whether
      or not he was himself to be blamed for the injury was irrelevant, the only ex-
      ception being injuries deliberately caused. The compensation was fixed,
      meaning that it was limited to 50% of the previous wage[398] and that moral and
      material damage were excluded. In accordance to the *risque professionnel* the-
      ory,[399] this compensation had to be paid exclusively by the employer, irrespec-
      tive of his personal responsibility. The employer could however exonerate
      from this obligation by concluding an industrial injuries insurance with an in-
      surer of his choice or with the *Algemene Spaar- en Lijfrente Kas* (a State in-
      surance fund, ASLK).[400] The employee-victim was further protected against
      an insolvent debtor by the requirement that insurers needed authorization and
      by the establishment of a *Garantiefonds* (guarantee fund) – feeded by contri-
      butions of non-insured employers – that would pay the legally fixed benefits in
      case of insolvency of the employer.

140   The possibility of pressing civil charges against the employer was simply
      eliminated, except for the case that the employer had caused the accident in-
      tentionally. In this way, the arrangement could be presented as a just compro-
      mise that served the goal of social peace. The risk of all industrial injuries
      would be equally divided on the two components of industrial production:
      capital and work.[401] Each party would thereby gain by making a sacrifice. The
      employee could not obtain complete compensation but would in any case get
      some; the employer accepted to compensate in cases where he was not to
      blame but was in turn liberated from legal proceedings and uncertain damages
      to pay.[402]

141   Still, not everyone was convinced that the proposed transaction was balanced.
      In a minory note Denis defended the position that the Act should offer the vic-
      tim complete compensation because ultimately it was not only on employers
      and employees that the burden of industrial accidents should fall but on soci-

---

[396] Y. Mondelaers (*supra* fn. 371), 23–29. This author also mentions medical practitioners as a
      group in favour of a new compensation policy as they hoped to be (better) paid for their ser-
      vices (p.31).
[397] To the relief of the victim the Act contained the presumption that any injury that occurred dur-
      ing the execution of the labour contract, was also caused by it. The employer had the possibil-
      ity to prove the contrary.
[398] After a waiting periode of one week with dating back to the day after the accident.
[399] *Parl. St.* Kamer 1901–1902, no. 302, 4 and 43–44.
[400] This latter possibility was considered necessary to make insurance more accessible, to pro-
      voke competition and to prevent artificially high premiums. *Parl. St.* Kamer 1900–1901, no.
      123, 6.
[401] *Parl. St.* Kamer 1900–1901, no. 123, 3; M. Demeur, [1908] Pand. Belges, 140; F. Tanghe
      [1989] RdW 2, 67.
[402] *Parl. St.* Kamer 1901–1902, no. 302, 5, 6 and 44.

ety as a whole.[403] This could be realized by counting the premiums for full compensation in the prices of goods. The argument, particularly made by Trasenster in another minority note,[404] that the financial burden on industry and its international competitiveness would be to heavy,[405] was rejected by Denis under reference to the facts in Germany. And fair competition within the country could be maintained by making the insurance for work-related injuries compulsory.[406]

Compulsory insurance was in effect another and even more important point of contention. According to a parliamentary minority represented by Denis, a compulsory insurance realized by authorized mutual funds or by state fund would have many advantages besides the one already mentioned: There would be a maximum spread of risk; it would be the cheapest solution all the much so because of the absence of business profit[407]; it would provide the victim absolute security of compensation and offer the employer protection against complete ruin or bankruptcy; it was the only legitimate way of putting the burden of industrial injuries on production; and, with reference to German evidence, Denis argued that those mutual funds had a beneficial effect on prevention and medical care.[408]   142

Although one seemed to realize that compulsory insurance was indeed the most appropriate solution for securing the effective payment of benefits,[409] and although it was hoped[410] that the possibility of insuring would be generally applied, the AOW ultimately did not oblige employers to insure. Apparently, compulsory (even non-state!) insurance represented a level of state interference that could not be tolerated in the country that at the time counted as a paragon of liberalism.[411] Officially, the compulsory character of the insurance was not maintained because the controversy about it would have retarded the adoption of the whole Act whereas the risk of insolvency was ultimately considered a second order problem.[412] In parliament it was argued that an obliga-   143

---

[403] *Parl. St.* Kamer 1901–1902, no. 302, XII–XIII.

[404] *Parl. St.* Kamer 1901–1902, no. 302, XLIX.

[405] This concern was also made in the explanatory memorandum, *Parl. St.* Kamer 1900–1901, no. 123, 9.

[406] *Parl. St.* Kamer 1901–1902, no. 302, XIII–XV.

[407] Reference was made here to the excessive high premiums in Britain.

[408] *Parl. St.* Kamer 1901–1902, no. 302, I–X.

[409] In fact, the establishment of the *Garantiefonds* can be seen as a second best solution. That the arrangement chosen in the Bill might be theoretically less complete, was also admitted in the explanatory memorandum and the possibility of future changes – if experience would demonstrate the necessity thereof – was explicitly mentioned. *Parl. St.* Kamer 1900–1901, no. 123, 3.

[410] *Parl. St.* Kamer 1900–1901, no. 123, 5.

[411] A. Beeckman [1886] JT 367, 1173; J. Van Steenberge, De Arbeidsongevallenwet: tussen de 19e en de 21ste eeuw, [1990] *Belgisch Tijdschrift voor Sociale Zekerheid* (BTSZ) 23, 364. By way of defence reference was also made to similar arrangements without compulsory insurance in France and England, countries that demonstrated the same preference for the liberal tradition and whose markets were important for Belgian industry: *Parl. St.* Kamer 1900–1901, no. 123, 3.

[412] *Parl. St.* Kamer 1900–1901, no. 123, 3.

tion to insure could only be effective in case an elaborated state supervision would be provided for, the administrative costs of which were considered to elevated.[413]

144   Another concern was prevention. The possible loss of safety incentives was invoked by those who opposed the automatic compensation regime and for the reason of prevention wished to preserve civil liability law.[414] This argument was also brought up in relation to the possibility of insurance, but put aside with reference to the possibilities of premium differentiation and recourse.[415] The prevention issue also arose when the question was disputed whether the Act should offer compensation in case an injury was caused by a serious fault of the employer or employee. Referring to German and Austrian figures the negative effect on prevention was denied. Moreover, excluding this category of accidents was not expected to have a deterrent effect since the victim would be anyway compensated by means of social assistance or charity. And regardless of the compensation provided for by the Act, the faulty behaviour could still be sanctioned by criminal law and safety legislation.[416] At the insistence of employers, who in exchange for the sacrifices they had made wanted absolute security with regard to the absence of legal proceedings,[417] no distinction was made between normal and serious faults.

145   Since the AOW 1903 was made up as a compromise none of the interested parties was fully satisfied with the result that was reached. On the side of the employees the benefits accorded by the Act were judged insufficient and the repartition into halves of the *risque professionnel* was difficult to accept. Employers protested against the newly imposed expenditure that would threaten industrial prosperity. And some lawyers remained feeling uncomfortable with the 'impure' legal principles of the Act.[418] The vote in parliament resembled this reception. In none of the two legislative chambers a negative vote was noted since all agreed that the Act implied a clear amelioration of the employees' situation and one realized that no delay could be afforded. Still, a lot of parliamentarians abstained from voting and made critical remarks. It was expected that the Act would be temporary or transitory and would be in need of amending in the near future.[419]

3.4.2 A Second Shift?

146   In the decades that followed the AOW 1903 was indeed frequently modified, but these changes remained modest and mostly related to the level of compensation and the extension of the range of application. In 1941 for example, the

---

[413] *Parl. St.* Kamer 1901–1902, no. 302, 138–143.
[414] F. Tanghe [1989] RdW 2, 65–66.
[415] *Parl. St.* Kamer 1901–1902, no. 302, 59.
[416] *Parl. St.* Kamer 1901–1902, no. 302, 53–57.
[417] *Parl. St.* Kamer 1901–1902, no. 302, 45–46 and 48–50.
[418] M. Demeur [1908] Pand. Belges, 124x.
[419] M. Demeur [1908] Pand. Belges, 125–126 and 136–137; *Hand.* Kamer 1902–1903, 16 July 1903, 2006–2007; *Hand.* Senaat 1903–1904, 17 December 1903, 130.

war conditions led to the compensation of commuting accidents, a measure that after the war was confirmed by the *Besluitwet* (Act) of 13 December 1945. Two changes are nevertheless worth of being looked at in more detail. In 1971 the AOW 1903 was replaced by a new *Arbeidsongevallenwet* (Industrial Injuries Insurance Act, AOW 1971) and made the insurance against work-related injuries compulsory. But also in 1951 an important modifying Act was voted.

The Wet (Act) of 10 July 1951 that modified the existing AOW 1903 at first   147
sight appears modest in the scope and amount of changes, but to my opinion nevertheless marks an important step in the evolution of Belgian industrial injuries compensation. The Act indeed did not imply any structural adjustments[420] but brought about a number of changes[421] at the benefit of industrial injuries victims and their claimants. Possible changes with regard to the system of insurance, prevention or rehabilitation were explicitly postponed.[422]

As motivation for the proposed modification the explanatory memorandum   148
mentioned: 'the desire to respect and safeguard the high dignity of the workmen', 'to strengthen the spirit of peace and cooperation in the relation between employer and employees', and 'to promote social progress'.[423] As in 1903, appeasement through a better compensation seemed to be the prior aim. In fact, the legislative initiative had again been triggered by social unrest:[424] There had been wildcat strikes in the port of Antwerp and a general strike in July 1950.

After the first World War it had become clear that employees weren't able to   149
bear their 50% of the *risque professionnel*, and the demand was made for full compensation. A first adjustment was made in 1930 when a distinction was made between permanent and temporary disability. During the first 28 days of the disability the compensation would still be limited to 50%, but afterwards it would raise to 66%. As the compensation stayed low in comparison with unemployment and (starting from 1945) with sickness benefits, the dissatisfaction remained as well. The modifications made by the 1951 Act had to remedy this.

The Act raised the compensation to 80% for the first 28 days, and to 90% af-   150
terwards. Once the disability was consolidated, the compensation would reach 100%. An extra 50% was possible when the disabled person needed help from a third party. A lot of parliamentarians were disappointed that full compensation was not provided for from the first day on, but a majority followed the ar-

---

[420] Report-Humblet, Pas., 712.
[421] See the enumeration of these by the competent Minister himself: *Hand.* Senaat 1950–1951, 27 June 1951, 1697.
[422] *Hand.* Kamer 1950–1951, 22 May 1951, 20.
[423] Explanatory memorandum, p. 2. (My translation.)
[424] *Hand.* Kamer 1950–1951, 22 May 1951, 4.

gument that this would hinder the reintegration to work that was considered necessary economically as well as morally.

151   More important and interesting than the increase of the level of benefits is the argumentation put forward to legitimate it, since the general consensus that existed with regard to the former, was absent in relation to the latter. The explanatory memorandum was very concise at this point. After stating that the 1903 scheme had served workmen well, it made clear that the division of responsibility that the existing scheme implied, had to be judged inequitable taking into account the progress made by social insurances and had to be replaced by the 'more human regime' of full compensation.[425]

152   The rapporteur of the *Kamer van Volksvertegenwoordigers* (the lower chamber, Kamer), Humblet, for his part, tried to demonstrate[426] that the new principle of full compensation related to the civil law principles already accepted before 1903. By consequence, although this new principle brought about a fundamental transformation of the legal foundations of the AOW, it did not constitute a legal revolution. It was only considered the last stage of the enterprise started in 1903. Whereas the AOW 1903 originally divided the risk of work-related injuries over the two central forces in the enterprise, labour and capital, the victim of an injury would from now on bear no longer any risk. The exclusiveness of the employer's responsibility was justified by the nature of the labour contract itself. As jurisprudence before 1903[427] had made clear and as had been confirmed by article 11[428] of the *Arbeidsovereenkomstenwet* (Act relating to Labour Contracts, WAO) of 10 May 1900, it was the simple fact of the labour contract itself that rendered the employer debtor of security. What the 1951 Act added to this interpretation was the reversal of the burden of proof as suggested by Sainctelette but rejected by the Hof in 1889. Henceforth, the employer would be presumed responsible for any accident that occurred during and due to the execution of the labour contract unless he could prove otherwise. In the eyes of Humblet, the new Act had the merit of finally giving article 11 of the WAO its full legal and social range.

153   Parliamentarians that wanted to express the employers' stand, heavily criticized this reasoning. They argued that the presumption of the employer's responsibility did not stroke with reality, as a recent survey of the International Labour Organisation (ILO) had shown that employers could only be held responsible for one fourth of the total number of accidents. Nor did the argu-

---

[425] Explanatory memorandum, p. 1. With full compensation was meant a compensation equal to the wage lost due to the accident.

[426] Report-Humblet, *Pas.*, 712–713.

[427] A jurisprudence which was however not followed by the *Hof van Cassatie* (The judgment of 28 March 1889, Pasic., 1889, I, 162), cf. *supra* no. 133.

[428] This article recognised a limited contractual duty of safety for the employer and stipulated that the employer had to establish decent working conditions and had to render the employee the first aid in case of an accident. Although this obligation did not extend to a duty to compensate, it still constituted a clear reference to the theory of Sainctelette.

mentation given by Humblet explain why commuting accidents had to be compensated for. Moreover, sickness and unemployment were not fully compensated either.[429]

But also other parliamentarians expressed the feeling that the legitimation offered by Humblet was somewhat farfetched.[430] Instead of the 'ingenious' way in which Humblet had tried to give the modifications made by the 1951 Act a legal foundation, based on the traditional principles of civil law, they preferred to admit that these modifications were sufficiently justified by the remarkable evolution of social circumstances since 1903. Or, as a member of parliament stated with regard to the new principle of full compensation: 'Dans l'état actuel de la legislation sociale, c'est une notion qui s'impose et qu'aucune considération juridique ne pourrait valablement contredire'.[431]   154

When reading the parliamentary discussions, one indeed gets the impression   155
that a general atmosphere of 'social generosity' made possible by the favourable economic situation and the balance of power in parliament, rather than legal considerations were at the basis of the enacted improvements.[432] That the realization of an ever more progressive social legislation was a matter of prestige between the Western European countries also seemed to have played a role. In that respect the Act of 1951 was intended as a piece of legislation to be proud off.[433] What seems then to explain the effort made by Humblet is a certain embarrassment towards the fast pace in the development of social legislation and the need, typical of lawyers, to find a connection with existing (and accepted) legal principles rather than having to use the *ultima ratio* of the legislator's will.[434]

In 1971 a further step was set. The AOW 1903 was then replaced by the AOW   156
1971, which is at the moment still in force. The Act had the purpose of coordinating and modernizing the compensation scheme along the lines of the recommendations made by the *Nationale Arbeidsraad* (National Labour Council, NAR). Important changes were: equalizing the range of application of the Act with that of the social security scheme; a broad and flexible definition of the concept 'industrial accident'; increasing the benefit in case of temporary disability from 80 to 90%; the obligation for the insurer to pay the legally fixed benefits right away even when a civil proceedings is installed; and, of course, making the insurance for industrial injuries compulsory.

---

[429] *Hand.* Kamer 1950–1951, 22 May 1951, 9–10 and 14–16.
[430] *Hand.* Senaat 1950–1951, 27 June 1951, 1694–1695.
[431] *Hand.* Kamer 1950–1951, 22 May 1951, 9. In the same sense: *Hand.* Senaat 1950–1951, 27 June 1951, 1699.
[432] *Hand.* Senaat 1950–1951, 27 June 1951, 1694–1695; *Hand.* Kamer 1950–1951, 22 May 1951, 14.
[433] *Hand.* Senaat 1950–1951, 27 June 1951, 1697 and 1699.
[434] See Humblet's reaction to the critic on his legal foundation (*Hand.* Kamer 1950–1951, 22 May 1951, 10) and the statement of the competent Minister (*Hand.* Senaat 1950–1951, 27 June 1951, 1697).

157   The motivation of this latter measure was found in the fact that Belgium was
        the only EEC-country without such an obligation and the consideration that
        compulsory subjection to social legislation was general. Moreover, in practice
        not much would change since most employers already were insured.[435] The
        compulsory character of the insurance was seen as the next step in an evolu-
        tion in which the Acts of 1903 and 1951 were the former stages. To the gov-
        ernment's opinion it was time to have the *risque professionnel* notion disap-
        peared and to bring the legislation with regard to work-related injuries closer
        to this of social security. A new foundation had to be found in the general
        thought of social security legislation: the need to protect the employer and his
        claimants as much as possible against the consequences of events that put him
        in the impossibility to provide for himself and his family. As a consequence,
        the employer would no longer be responsible for the suffered damage (as un-
        der the 1951 Act), but would only have one obligation left, to conclude an in-
        surance.[436]

158   It was the *Raad van State* (Council of State, Raad) that stressed the profound
        influence of the insurance obligation on the legal nature of the compensation
        scheme. Henceforth, it was no longer the employer but his insurer or the
        Fonds voor Arbeidsongevallen (Industrial Injuries Fund, FAO) that were the
        debtor of the legally fixed benefits. The employee had no longer a direct claim
        on the employer, and this latter had no other obligation than to pay an insur-
        ance premium. The insurance itself did not cover any longer the risk of the
        employer to pay these benefits, but the risk of the employee to suffer a loss of
        wage and to be in need of health care as a consequence of an industrial injury.
        This meant a change from a liability insurance into an accident insurance.[437]

159   According to the Raad all these changes made the AOW 1971 more similar to
        the compensation schemes of social security. As differences with social secu-
        rity schemes that remained, the Raad mentioned the fact that unlike the bene-
        fits, the premium of the insurance was not fixed by government but by agree-
        ment between the employer and the insurer, and the fact that the employer had
        the obligation to conclude a private insurance instead of having to join a social
        security agency. With regard to this latter difference, the Raad moreover made
        the remark that a further rapprochement to the social security scheme would
        require the possibility for the employer to conclude an insurance with the
        FAO.[438]

160   The Raad also made some considerations regarding the industrial preference.
        The Raad made mention of the fact that in general the benefits of the AOW
        1971 were more important than those of social security, and that this differ-
        ence could not be maintained in the future if one departed from the principle

---

[435] *Parl. St.* Senaat 1969–1970, no. 328, 29.
[436] *Parl. St.* Senaat 1970–1971, no. 215, 3–4 and 6.
[437] *Parl. St.* Senaat 1969–1970, no. 328, 69–71.
[438] *Parl. St.* Senaat 1969–1970, no. 328, 70–72.

that both schemes had the same object in view.[439] Remarkable, but also typical of the already mentioned atmosphere of social generosity, is that the Raad recommended to adjust the provisions in the AOW 1971 that were less favourable than those of the social security scheme[440] instead of the other way around, thereby in fact reinforcing the industrial preference as long as a complete unification with social security would not have been realized. Whether a further unification should be considered, was a question the Minister was asked to submit to the NAR.[441]

3.4.3 Recent Developments

Since the introduction of the AOW 1971 no major changes have taken place in Belgian policy with regard to work-related injuries.[442] The compensation scheme developed in 1903 and refined in 1971 has proven to be a stable compromise between employers, employees and the state. None of the interested parties has since then questioned the basic assumptions of the system that seems to work to everyone's satisfaction.[443] Legislative changes in the past thirty years were more technical than principal in character. How work-related injuries can best be compensated is in contemporary Belgium no longer a political issue.[444]

161

Some changes however are worth being mentioned because they point to an increased awareness of the importance of prevention as a policy goal. In 1987 the establishment of a *Technisch Comité voor Preventie* (Technical Committee for Prevention, TCP) in the FAO was envisaged. This TCP has the task to stimulate and organise research with regard to the prevention of work-related injuries and is asked to formulate recommendations in that field.

162

Other changes regard the employers. By a *Wet* (Act) of 1999 the list of cases in which the employer loses the advantage of his immunity to civil actions has been extended by the case where the employer has acted against the written prescriptions of the safety inspectorate. And since a *Wet* (Act) of 2003 the employer is also obligated to report the 'severe' industrial accidents to the safety officials who then can decide to appoint an expert with the task to investigate the causes and circumstances of the accidents and to make recommendations on how to avoid repetitions.

163

---

[439] *Parl. St.* Senaat 1969–1970, no. 328, 72–73.
[440] *Parl. St.* Senaat 1969–1970, no. 328, 74.
[441] *Parl. St.* Senaat 1970–1971, no. 215, p. 37.
[442] By the *Wet* (Act) of 28 June 1981 the AOW was formally integrated in the Social Security scheme.
[443] With exception maybe of the private insurers who complain that the industrial injuries insurance is not profitable enough: W. Rauws, Financiering van schade veroorzaakt door arbeidsongevallen en (nieuwe) beroepsziekten: België als wenkend voorbeeld?, in: M. Faure, T. Hartlief, *Schade door arbeidsongevallen en nieuwe beroepsziekten* (2001), 112.
[444] W. Rauws (*supra* fn. 443), 118.

164   The extension of cases in which the employer loses his civil immunity[445] has
        been evaluated by some (together with the successive rises of the benefits, cf.
        supra no. 149 and 150) as cracks in the historical compromise of 1903 and as
        a possible (jurisprudential) evolution towards a restoration of the employer's
        civil liability,[446] but until now no further legislative interventions have fol-
        lowed.

4. Evaluation

4.1   Approach

165   In the former chapter an overview has been given of the different consider-
        ations and motives that have been explicitly put forward during the parliamen-
        tary proceedings or that were mentioned in the literature. For evaluating these
        considerations the following questions seem important to ask. First, what was
        the essence of the shift noticed and how do the changes relate to the discerned
        functions of compensation systems? Second, which were the main arguments
        for the shift and how do these arguments take into account the differences and
        typicalities of compensation systems with regard to their possible functions?
        As mentioned before (cf. *supra* no. 2), to be able to produce a critical evalua-
        tion and comparison of the various shifts, it is important to know to what ex-
        tent certain modifications were deliberately pursued and, even more impor-
        tantly, to find out which of the consequences connected with or stemming
        (naturally) from these modifications were envisaged by the government and
        which were not. This evaluation will be the content of section 4.3.

166   As Ogus[447] has rightly pointed to in his contribution to this research project,
        this kind of *public interest analysis* constitutes only one perspective or frame-
        work to explain policy developments. It would indeed be naïve to believe that
        policies and laws simply follow from a rational search of a solution to an ob-
        served problem made in the public interest. Policy choices may indeed equally
        reflect the private interests of groups that were able to influence the law-mak-
        ing process to their benefit. The fact that certain pressure groups have suc-
        ceeded in having their own interests defended or confirmed by the principles
        and procedures chosen for, might even provide an explanation for the ob-
        served inconsistencies in shifting from one compensation system to the other.
        An attempt to such a *private interest analysis* will be made in section 4.4.

167   Although I have started (cf. *supra* chapter 2) from a fairly clear opposition be-
        tween a private law compensation system and a public law compensation sys-
        tem, in practice, it is not easy to decide whether or when a shift between pub-

---

[445] In 1964 resp. 1998 the immunity had already been lifted with regard to commuting accidents
    and traffic accidents. See: V. Vervliet (*supra* fn. 368), 174 *et seq.*
[446] V. Vervliet (*supra* fn. 368), 293–306.
[447] A. Ogus, Shifts in Governance for Compensation to Damage: A Framework for Analysis, in:
    W.H. van Boom/M.G. Faure (eds.), *Shifts in Compensation between Private and Public Sys-
    tems* (2007).

lic law compensation and private law compensation schemes (as defined in our research project) has occurred. The transition from one to the other appears to be hardly ever clear-cut, but on the contrary gradual or incomplete. Illustrative is the case of Belgium. Although at no point Belgium seems to have experienced a real 'shift', it is undeniable that the compensation scheme at present is all but the same of that before 1903 and displays much more features of a public law compensation scheme.

Still, as the previous chapter and the chapter by Engelhard show, it is possible   168
to group the policy changes in compensating for work-related injuries together in three time periods which appear to be by and large the same for all the examined countries. From this follows that, although not all countries may have experienced a real shift in compensation in each period, there seems to have existed at certain moments in time some shared and more general social circumstances that *could* provoke a decision to consider policy changes. Therefore, it appears useful to first briefly sketch this shared contexts in which changes and shifts have occurred.

4.2   General Background for the Shifts

In most countries the establishment of a compensation arrangement for work-   169
related injuries constituted the first expression of what eventually and gradually would become a state-organised overall social security system. It seems that this 'pioneer system of social security'[448] continues to offer a rather representative reflection of the general evolution and tendencies in social security policy as a whole. My research and that of Engelhard allows to discern an evolution in which three periods of time can be distinguished, an evolution that returns in studies where a general overview of social security policy is aimed at.

The first phase, situated at the end of the 19th century, is the one where the   170
first initiatives are taken, where a general awareness arises that society organised as a state needs 'to do something'. An immediate cause for action was delivered by the so-called 'socialist danger', i.e. the (threatening) rise of the labour movement both in the street and in parliament. Policy initiatives can then be seen as an attempt to maintain the new industrial context by meeting to some extent the complaints of the workers who were faced with the negative effects of the industrialization. Social appeasement would become an important policy goal. Especially in Belgium and Germany, legal initiatives followed close on social riots. It is no surprise then that at least formally[449] and theoretically[450] the responsibility of the employer was accentuated. Since both the content of the established compensation schemes as the fact of state intervention in this area were at the time considered as rather revolutionary, the dif-

---

[448] W. Beveridge (*supra* fn. 261), 156.
[449] In fact, employees took their part of the financial burden since they mostly lost the possibility of full compensation.
[450] Cf. the theory of *risque professionnel.*

ferent initiatives in this phase are accompanied by elaborate and explicit motivations with regard to the arrangement chosen, its legal foundations, underlying concepts of justice, goals to aim at etc.

171   A second phase commences after the second World War, when the war events have brought about a reconfirmed and more strongly felt national solidarity, and economic prosperity gradually allows 'to do more'. It is a period of euphoria and social generosity where the compensation schemes are further developed (often at the instigation of governments themselves with a great personality as a moving force) but without much explicit motivation or with a motivation that is more ideological in character. The urge to internationally distinguish oneself as a model welfare state also plays an important role. Notable for this period is that the industrialization is no longer looked upon as a process or situation brought into being by one specific category of people (who consequently have to carry the burdens thereof), but as a social (and irreversible) fact, the advantages and disadvantages of which have to come down to society as a whole.

172   The last phase that started from the late 1970s on and still seems to be going, is the period of critical reflection and interrogation where one openly asks 'haven't we done too much?'. Despite of the profoundness of the question, policy reflections are rather superficial and hasty and are dominated by considerations of expenditure reduction and cost containment. Policy changes usually come down to taking a step backward by reducing the scope of the regime, downsizing the benefit level, or shifting costs to the private sector.

173   As mentioned earlier it is remarkable that, irrespective of the compensation systems they have adopted and whether this has resulted in a shift in governance or not, all of the examined countries show this evolution and at about the same point in time. From this conclusion follows that the felt need to make changes to the compensation scheme in place is not exclusively (or maybe not even primarily) triggered off by the as deficiencies perceived characteristics of that scheme but by more universal external factors.

4.3   Public Interest Analysis

174   In all examined countries a change in the way of compensating for work-related injuries can be distinguished around the turn of the 20th century. The underlying cause of this change is for all countries identical, that is the apparent inadequacy of the compensation system in place, being civil liability law, in the light of the changed socio-economic circumstances due to the process of industrialisation. The uttered dissatisfaction was general and regarded the heavy burden of proof, the slow, costly and adversarial character of the system and the uncertain and meagre outcome. Consequently, the goal to be achieved was crystal clear: a compensation scheme that was easier accessible and offered adequate compensation in more cases. In this respect a somewhat divergent position is taken by England where, initially at least, the primary aim was

to attain safer working conditions. It was not until Lord Chamberlain came to the fore that changes in the liability system were primarily motivated by the wish for better and easier compensation than as a means of stimulating employers in taking more and better precautionary measures.

The compensation schemes that subsequently were brought into being all had   175
one feature in common: They offered compensation irrespective of fault. Not the question who was to blame but the guarantee of compensation stood central. In Germany and the Netherlands this goal was reached by the introduction of a public law insurance duty for employers based on the *risque professionnel*. Allowing employers immunity for their civil liability compensated for this new obligation. In Belgium and England employers faced a strict liability to compensate. In all examined countries the financial burden of work-related injuries was thus more clearly put on the employer, an allocation away from the employees-victims[451] and from poor relief.

For Germany and the Netherlands this change can be evaluated as a real shift   176
since the system of protection obtains a collective and compulsory character, and civil liability law is largely put aside by adjudicating employers civil immunity. The new scheme more clearly reveals the features noted in the right column of the compensation diagram (cf. *supra*): low threshold, limited compensation and maximal spreading of losses. The choice for a compulsory insurance was also a deliberate one. It was motivated by reference to the possibility of loss distribution and the beneficial effects on social peace and securing compensation, and was thus consistent with the postulated objectives. In both countries there was also a clear awareness of the fact that with civil liability law a possible preventive effect would go lost. A solution was found in differentiating the premiums in function of the risk the companies represented with regard to the occurrence of industrial injuries.

Although the problems to be tackled and the goals to be achieved were formu-   177
lated in a similar way to those in Germany and the Netherlands, Belgium and England came to a different solution and remained closer to the existing liability law. The result here can be described as a 'socialized liability law'.[452] Their rejection of compulsory insurance was however based on different reasons. The English rejected it in principle, because they feared that it would be counterproductive to preventive efforts from the employers.[453] As noted earlier, obtaining safer working conditions was at least as important as a policy goal as producing an easier and secure compensation, and only a clear and exclusive liability of the employer was hold to be able to bring about that result. The Belgian rejection was of a different (more ideological and pragmatic) nature, and was not so much directed at the principle of insurance but at the compul-

---

[451] Be it only partially since employees were denied full compensation.
[452] B. Barentsen (*supra* fn. 68), 140.
[453] The same reason, but then directed at the employees, served as a justification for the low level of benefits. That a high level of benefits could act as an incentive for prevention on the part of the employers, was not taken into consideration.

sory character of it. The prevention argument was in any case not decisive as it was in England.

178   All in all the Belgian system ultimately more strongly resembled the German and Dutch situation than the English. There was the same adjudication of civil immunity and compensation security was indirectly realized through the *Garantiefonds*. The level of benefits however remained comparably low, and premium differentiation was in the hands of private insurers. Apart from the tension between the compensation and the prevention goal, the English scheme demonstrated also an inconsistency with regard to the litigation objective. To compensate for the low level of benefits and to make full compensation still attainable, the possibility of civil action was preserved. In fact, this way litigation was made an inherent feature of the scheme.

179   A second wave of important changes can be situated after the second World War. The motive for the changes was again the pursuit of more protection. In Germany and Belgium the alterations did not constitute a real 'shift', although this has to be nuanced for the case of Belgium. In Germany a further extension and 'socialization' of the *risque professionnel*[454] could be observed, as well as an increased attention for prevention and rehabilitation.

180   In Belgium, the dissatisfaction with regard to the low benefits of the AOW 1903 had continued to grow and eventually led to an amendment of the law in 1951 after the social riots the year before. Again, better compensation was supposed to bring social appeasement. The adoption of the so-called 'principle of full compensation' can however also be explained by my post-war social generosity argument (cf. *supra* no. 170). Revealing in this respect is the underpinning motivation. The somewhat forced attempt to legitimize the change under reference to the civil (and even contractual) liability of the employer, was heavily criticized and according to many ought to be replaced by the simple but sufficiently justifying argument that this new principle was in line with the evolution of the social circumstances. It is no surprise then that the introduction in 1971 of the compulsory character of the industrial injuries insurance – by which Belgium came another step closer to the (original) Dutch and German system – was seen as nothing more than a logic step in view of the 'general thought of social security' and could be motivated referring to foreign examples and the fact that 'all social insurances were compulsory'.

181   Looked at from the compensation diagram the new regime produces some inconsistencies. Given the (nearly) full compensation level and the absence of any employee's contribution or state subsidy, the risk of industrial injuries is entirely allocated to the employer,[455] which sits uneasy with the general princi-

---

[454] Cf. The '*unechte Unfallversicherung*' that in fact has come to comprise a part of the *risque social*.

[455] However, the employers can spread their risk due to the insurance possibility.

ples of a social security system, especially when this risk becomes more and more non work-related. Nor is the full compensation in conformity with the easy access to the scheme; the retention of the employers' civil immunity on the other hand is. As the Belgian *Raad van State* rightly indicated, it would also be more logic and in line with the compulsory character if, next to the possibility of private insurances, there was also the availability of taking out insurance with a public body. Nevertheless, the Belgian scheme as it was modified in 1971 is still in force and appears as yet to be able to realize the original goals of compensation security and industrial peace.

Real shifts have occurred in this period in the Netherlands and in England. In both countries compensation for work-related injuries was formally integrated in a more general social security system accompanied by liability law as a complementary protection. In the Netherlands work-related and non work-related risks were integrated in a new mandatory social insurance (thus based on the *risque social*) paid for by employers and employees, and civil liability law was reintroduced as a complementary scheme in order to obtain full compensation. For this latter risk there was no obligation to insure.[456] In England a compulsory social insurance based on the *risque professionnel*[457] was adopted, financed by employers, employees and the Exchequer. Also for the employers' civil liability, that remained as an alternative remedy, insurance became now compulsory.[458]   182

Remarkable is that this second Dutch shift did not follow from any dissatisfaction with the compensation scheme in place, as was the case for the first Dutch shift. On the contrary, the OW was frequently said to be 'a fairly satisfying' arrangement. The only motivation for this shift seems to be a general aspiration to establish an all-encompassing social security scheme in which adequate benefits are available to all needy citizens regardless of causality. The argument of the post-war euphoria and social generosity as an explanation for change finds a marked illustration here. The result is an over-emphasizing of the compensation goal, sometimes against one's better judgment, it seems. There was for instance a clear awareness of the fact that leaving the risk-relatedness of premiums would mean a loss of prevention, but still this alteration was put through because of its 'incompatibility with social insurance principles'. And although a (maybe compensating) preventive effect could be expected from the return of the employers' civil liability, this argument was not   183

---

[456] In a sense one could make mention here of a 'shift back' to the situation before the OW, where protection had to be found on an individual basis via liability law and without an insurance obligation.

[457] Although officially other reasons explain the retention of the industrial preference, there is general agreement about the fact that is simply politically not feasible to withdraw this workers' achievement.

[458] In both countries, the risk of having to pay a statutory defined level of compensation is thus reallocated from the employer to the community of employers and employees or to the community at large, while the risk of incomplete compensation is reallocated from the employees to the employer.

used. On the contrary, also the upheaval of the employers' civil immunity appears mainly to be motivated by the wish for more compensation. The possible danger of this revival of civil liability for the industrial peace was on the contrary waved aside. Nor was there any awareness of or concern about the fact that the new scheme tried to combine elements from one compensation system (i.c. the easy access of social security based on solidarity) with (conflicting) elements of the other (i.c. the full compensation linked to corrective justice and civil liability).[459]

184   The English shift on the other hand undoubtedly was a reaction to a general dissatisfaction with the compensation scheme in force, a dissatisfaction that was predictable taking into account the many inconsistencies of the WCA. Still, the serious deficiencies of the old scheme constituted only one of the factors that influenced the decision to undertake major reforms. Another important factor was the war and the new sense of solidarity that it had brought about, which made class and political differences for a time at least less relevant and caused a consensus on the fact that the country needed social security.[460]

185   In comparison with the former scheme, the new one was expected to offer secure and sufficient compensation and a lesser amount of litigation. Prevention appeared to have lost its primordiality as was also shown by the fact that Beveridge's special levy on hazardous industries was not accepted. Nor was prevention the first or main argument in favour of the retention of the employers' civil liability; the concern to avoid 'undercompensation' and the influence of the TUC seem to have been more decisive elements. The compulsory character of both the liability[461] and the social insurance was also motivated with reference to the desire of compensation security.

186   The current English compensation scheme for work-related injuries shows a lot of inconsistencies taking into account the postulated goals. As the Dutch scheme it tries to combine the easy access of a social insurance scheme and the full compensation of civil liability law. Contrary to the Dutch case however, this combination seems necessary given the relatively low level of social insurance benefits.[462] But as civil proceedings remain possible or even necessary, high amounts of litigation and administrative costs will inevitably remain part of the system as well.

---

[459] B. Barentsen (*supra* fn. 68), 179. It is interesting to observe that this violation, which finds its expression in the second shift, seems to be the result of an incomplete awareness of compensation logics during the previous shift. Since the civil liability immunity introduced in the first shift was not motivated by reference to the compensation logic, it may have been easier to come to an infringement of this logic in a later shift when reintroducing this liability.

[460] Cf. J.C. Brown (*supra* fn. 204), 21–22.

[461] With regard to the employers' liability compulsory insurance meant again sacrificing prevention for compensation.

[462] A lot of long-term disabled employees would have to make an appeal to social assistance: P. Cane (*supra* fn. 7), 280–281.

Although in all West-European countries the 1970s ushered in the crisis of the   187
welfare state with the subsequent economy measures, only in England and the
Netherlands this has affected the industrial injuries compensation scheme. In
both countries an ongoing search seems to be taking place for a new balance
between all interested parties in compensating work-related injuries. The right
to benefits has been more reserved for the 'more seriously disabled' and the
compensation for the first period of sickness has been (partially) privatized. In
general, costs have been gradually shifted towards the employers and their insurance companies, irrespective of the fact that insurability problems seem to
pop up.[463]

4.4   Private Interest Analysis

The direction and content of policy changes are not only the result of rational   188
analysis and subsequent made choices of goals and means but also reflect the
way in which certain groups in society have been successful in having their
private interests protected and defended. With regard to the compensation of
work-related injuries the potential influence of employers and employees in
this respect is obvious. As became clear in the previous chapters the legislation and regulation in this field could most of the time be considered as attempts to reach a compromise between the private interests of these two important groups in industrial society. But also other groups have had their
impact.

Obviously, the role of employees was indispensable if only for the mere   189
awareness of the fact that the compensation of work-related injuries constituted a social problem to be tackled. In England, they even succeeded in domi-
nating the statement of the problem and having it defined as a lack of safety
instead of compensation. The pressure of employees grew into influence once
they got organised by means of trade unions and socialist political parties and
qualified to vote. In later developments they mostly would be successful in increasing and expanding the provided benefits and having their 'acquired
rights', such as the industrial preference, preserved.

Employers especially seem to have had an impact in the first developments at   190
the end of the 19th century. After they could no longer successfully prevent
government action in this field, they were able to channel this new government endeavour away from a threatening widening of their civil liability into
compensation schemes that were equally beneficial for themselves. They also
succeeded in having the significance for the nation's economy of their international competitiveness accepted as an important and ever recurring argument
in parliamentary discussions, in this way mitigating the laying down of extra
costs on them.

---

[463] See *supra* no. 129 and for the Netherlands: M.G. Faure/T. Hartlief (eds.), *Verzekering en de groeiende aansprakelijkheidslast* (1995); M.G. Faure/T. Hartlief (*supra* fn. 2).

191   Since the first initiatives with regard to a statutory organised compensation scheme for industrial injuries coincided with the emergence of the insurance business, insurers too became an interested party. They sensed a new lucrative market and became the allies of the employers against compulsory state insurances, in most countries not without success.

192   Universal suffrage made political parties interested in the workman's vote and thus susceptible to reforms to their advantage. In order to defend their parliamentary positions of power, even non-socialist traditional parties often took the lead in the search for ameliorations, such as the British Conservatives that enacted the WCA.

193   Also lawyers played their role. At the end of the 19th century in all examined countries a form of legal conservatism could be noticed. To protect the dogmas of civil law they objected to solutions within the existing civil law rules and in this way contributed to the establishment of a new 'social' law. The role of the judiciary however differed: While in Belgium and the Netherlands the judges gradually interpreted the law to the advantage of the victims-employees (as seems again to be the case in the most recent developments) and stimulated in this way legislative action, they tried to obstruct legal developments in England. First, by introducing three legal fictions known as the 'Holy Trinity', later by a narrow interpretation of the Employers' Liability Act and the acceptance of contracting out. Eventually, also in England judges would plea for an insurance system in order to protect the integrity of the common law.

194   Finally, even the government in place itself can be seen as a party with own interests to defend. Since budgets are essential to governing, governments can have an interest in shifting or simply limiting expenses. At the end of the 19th century, the shifting of costs away from social assistance or poor relief clearly constituted a factor in establishing a new industrial injuries compensation scheme. And in actual developments 'cost containment' seems to be an independent policy goal.

4.5   Final Conclusions

195   What final conclusions can be drawn from this research with regard to the shifts in compensating work-related injuries or even with regard to shifts in compensation in general? Let me first repeat the conclusions already mentioned earlier.

196   It appeared that shifts hardly ever stem from the features themselves of the compensation scheme in place, but are primarily triggered off by more general socio-economic circumstances.[464] The perceived possibilities (financially and other) by policy makers seem to influence the direction of the shift: Economic prosperity and a belief in (the necessity of) government intervention result in more public law based schemes, while periods of economic depression show a retreat of government and a shift towards private actors. In this sense, changes

---

[464] Cf. *supra* no. 173.

in the compensation of work-related injuries reflect quite well more general changes in the global social security policy.[465]

Shifts also appeared to be hardly ever clear-cut, but on the contrary often grad-   197
ual, incomplete or even inconsistent.[466] Although there seems to be a general awareness of the typicalities of each compensation system and of the advantages and disadvantages of each of them with regard to the different possible policy goals, in practice, compensation systems are seldom encountered in their pure theoretical form. The need to make political compromises, the occurrence of less well considered *ad hoc*-modifications to the regulations in place, and the successful pursuance of private interests by pressure groups, can offer an explanation for this.

Whether and when shifts have been successful or not, is a question that cannot   198
be addressed in general, since the answer mainly depends on the relative importance of the different policy goals pursued. Logically, it is neither possible then to define the compensation scheme that has been most successful in compensating work-related injuries. Still, when one takes the pursuance of peaceful industrial relations together with a secure and sufficient level of compensation as the *leitmotiv* in the history of compensating work-related injuries, one could suggest the bold thesis that some kind of 'ground model' seems to impose itself as the least contested or the most satisfactory in the sense that none of the interested parties has made urgent demands for increased benefits or other profound adjustments.

This compensation scheme demonstrates the following features: 199

- a compulsory insurance based on the *risque professionnel*;
- the financial burden put fully upon the employer;
- the accordance of civil immunity to the employer;
- statutory defined level of earnings-related benefits varying from 66% up to 100%;
- a possibility of premium differentiation.

A compensation scheme with these features has been in force for a long time   200
now in Germany and Belgium. As we have seen, in neither of these two countries the basic assumptions upon which the system is based, have been seriously questioned since its introduction. And although one has to be extremely careful in establishing causal relations, these countries also score well with regard to their accident rates.[467]

---

[465] Cf. *supra* no. 169.
[466] Cf. *supra* no. 167.
[467] Cf. the contribution of Philipsen. It could be an interesting idea for further empirical analysis to compare the results with regard to accident rates, prevention etc. between compensation schemes that do and schemes that do not demonstrate the characteristics of the ground model in order to understand whether this model only appears to be the best compromise between competing private interests or is also best able to realise the public interests.

	1st period	2nd period	3rd period
NL	*OW 1901* Compulsory insurance (based on risque professionnel) Benefits: 70% *(later: 80%)* Civil immunity Financial burden: employer	*WAO 1967* Social insurance (risque social) Benefits: 80% *(from 1985: 70%)* Civil action as a supplement Financial burden: employer + employee	*WIA 2006* Social insurance (risque social) Benefits: 70% *(for fully disabled)* Civil action as a supplement Financial burden: employer (individual premium differentiation)
B	*AOW 1903* Employers' liability to compensate (=strict liability) Benefits: 50% Civil immunity Financial burden: employer	*AOW 1971* Compulsory insurance Benefits: 90-100% Civil immunity Financial burden: employer	
E	*WCA 1897* Employers' liability to compensate (=strict liability) (contracting out possible) Benefits: 50% Civil action possible, but no cumulation Financial burden: employer	*IIS 1946* Social insurance (risque professionnel) Benefits: not earnings-related Limited cumulation with civil action Financial burden: employer + employee + Exchequer 1969: compulsory liability insurance 1989: recoupment scheme	
G	*UV 1884* Compulsory insurance (based on risque professionnel) Benefits: 66% Civil immunity Financial burden: employer	*UVNG 1963*	

201   The two countries were one has not yet been able to find a definite or stable equilibrium with regard to the compensation of industrial injuries on the other hand (England and the Netherlands), currently have schemes that are most removed from this model. However, in the Netherlands a similar scheme has been in force from 1901 up to 1967 and was equally said to have been working 'fairly satisfying'. In this respect it is maybe not that surprising that the

Netherlands, where the actual compensation scheme for industrial injuries does not demonstrate the abovementioned features, was in its search for adjustments somehow tempted to return to this system when considering the reintroduction of an extra provision for the victims of industrial injuries and occupational diseases based on a mandatory *risque professionnel* insurance.

Of course, it would be interesting to know why this model appears to be the least contested. Does the absence of real contestation follows from the listed features (and if so, which of them then play a determining role in this?)[468] or is it the mere result of sheer *inertia* of the interested parties? Moreover, the fact that this ground model seems to be successful in establishing social appeasement, says nothing about its fitness with regard to the realisation of the enumerated policy goals (cf. *supra* no. 6). In other words, this ground model may appear to be the best compromise between different private interests, but does it also succeed best in realising (a combination of) the different policy goals put forward from the public interest perspective? Only further research (cf. fn. 467) and the evolving of time itself will deliver an answer to these questions.

202

Bibliographic References

BARENTSEN, B., *Arbeidsongeschiktheid. Aansprakelijkheid, bescherming en compensatie*, Deventer: Kluwer, 2003, 280 p.

BARTRIP, P.W.J., *Workmen's Compensation in Twentieth Century Britain. Law, History and Social Policy*, Aldershot: Avebury, 1987, 250 p.

BEECKMAN, A., Responsabilité des maitres et patrons. Les assurances contre les accidents du travail au Congrès de Liège, *Journal du Tribunaux*, 1886, n° 367, p. 1169–1174.

BEVERIDGE, W., *Social Insurance and Allied Services*, London: His Majesty's Stationery Office, 1942, Cmd. 6404, 299 p.

BROWN, J.C., *Disability Income. Part 1. Work-related injuries*, London: Policy Studies Institute, 1982, 317 p.

CANE, P., *Atiyah's Accidents, Compensation and the Law*, London: Butterworths, 1999.

COHEN, P., *The British System of Social Insurance*, London: Philip Allan, 1932, 278 p.

CORNISH, W.R. & CLARK, G., *Law and Society in England 1750–1950*, London: Sweet & Maxwell, 1989, 690 p.

DE L'HOPITAL, G., *L'assurance contre les accidents du travail en Allemagne*, Paris: Rousseau, 1904, 166 p.

DELVAUX, G., et al., *Honderd jaar sociaal recht in België*, 1987, p. 103–109.

DEMEUR, M., *Le Risque Professionnel. Traité théorique et pratique de la Loi du 24 décembre 1903 sur la Réparation des Dommages résultant des Accidents du Travail*, Extrait des « Pandectes Belges », Bruxelles: Larcier, 1908.

DEPREZ, U., et al., Van steun-en voorzorgskas tot fonds voor arbeidsongevallen, *Belgisch Tijdschrift voor Sociale Zekerheid*, jg. 23, 1990, nos. 6–7–8, p. 371–379.

EWALD, F., *L'Etat Providence*, Paris: Bernard Grasset, 1986, 608 p.

---

[468] It is also possible that it are not the features themselves but their balanced combination that constitutes the key to the success. This implies that a same satisfying equilibrium could also be reached by a combination of other characteristics. In other words, not the features of a compensation scheme but their balanced combination could prove to be the key element.

FAGNART, J.-L., Un régime nécessaire et exemplaire de réparation des atteintes à l'integrité physique. Rapport de synthèse, in FAGNART, J.-L. (ed.), 1903–2003. *Accidents du travail: cent ans d'indemnisation*, Bruxelles: Bruylant, 2003, p. 307–318.

FASE, W.J.P.M., De legitimering van het verplichtend karakter van de sociale verzekering, in: JASPERS A.Ph.C.M., et al. (eds.), *De gemeenschap is aansprakelijk ... Honderd jaar sociale verzekering 1901–2001*, Lelystad: Vermande, 2001, p. 47–67.

FAURE, M.G. & HARTLIEF, T. (eds.), *Verzekering en de groeiende aansprakelijkheidslast*, Deventer: Kluwer, 1995.

FAURE, M.G. & HARTLIEF, T., *Nieuwe risico's en vragen van aansprakelijkheid en verzekering*, Deventer: Kluwer, 2002.

FLUIT, P.S. & WILTHAGEN, A.C.J.M., Het risque social, in: JASPERS A.Ph.C.M., et al. (eds.), *De gemeenschap is aansprakelijk ... Honderd jaar sociale verzekering 1901–2001*, Lelystad: Vermande, 2001, p. 107–125.

GITTER, W., *Schadenausgleich im Arbeitsunfallrecht. Die soziale Unfallversicherung als Teil des allgemeinen Schadensrecht*, J.C.B. Mohr (Paul Siebeck): Tübingen, 1969, 289 p.

HARTLIEF, T., *Ieder draagt zijn eigen schade. Enige opmerkingen over de fundamenten van en ontwikkelingen in het aansprakelijkheidsrecht*, Deventer: Kluwer, 1997.

HENNOCK, E.P., *British Social Reforms and German Precedents. The Case of Social Insurance 1880–1914*, Oxford: Clarendon Press, 1987, 243 p.

HERMANS, P.C. & PRINS, R. (red.), *Causaliteit en arbeidsongeschiktheid: verslag van een studiedag over het 'risque professionel' in Nederland*, Zoetermeer: SVR, 1993.

WORK-RELATED INJURIES ADVISORY COUNCIL (IIAC), *The Work-Related Injuries Scheme and the Reform of Disability Income*, Position paper No. 5, 1990, 18 p.

JANSEN, C.J.H. & LOONSTRA, C.J., De personele werkingssfeer van de socialeverzekeringswetten 1900–1960, in: JASPERS A.Ph.C.M., et al. (eds.), *De gemeenschap is aansprakelijk ... Honderd jaar sociale verzekering 1901–2001*, Lelystad: Vermande, 2001, p. 89–105.

JASPERS, A.Ph.C.M., De politiek en de sociale verzekering, in: JASPERS A.Ph.C.M., et al. (eds.), *De gemeenschap is aansprakelijk ... Honderd jaar sociale verzekering 1901–2001*, Lelystad: Vermande, 2001, p. 25–45.

JONES, S., Social Security and Industrial Injury, in: HARRIS, N., *Social Security Law in Context*, London: Oxford University Press, 2000, p. 461–494.

KLEEBERG, J.M., From Strict Liability to Workers' Compensation: The Prussian Railroad Law, the German Liability Act, and the Introduction of Bismarck's Accident Insurance in Germany, 1838–1884, *Journal of International Law and Politics*, vol. 36, 2003, p. 53–132.

KLOSSE, S., *Menselijke schade: vergoeden of herstellen? De werking van (re)integratieregelingen voor gehandicapten in de Bondsrepubliek Duitsland en Nederland*, Antwerpen/Apeldoorn: Maklu, 1989.

KLOSSE, S., Van WAO naar WIA: een verantwoorde omslag?, *Sociaal Recht*, 2005, p. 247–257.

KLOSSE, S., Schadevergoeding via sociale zekerheid en aansprakelijkheidsrecht: communicerende vaten?, in: FAURE, M. & HARTLIEF, T. (eds.), *Schade door arbeidsongevallen en nieuwe beroepsziekten*, Den Haag: Boom Juridische Uitgevers, 2001, p. 1–18.

KLOSSE, S. & VONK, G., De betekenis van het recht voor de toekomst van de sociale zekerheid, in: KLOSSE, S. (ed.), *Sociale zekerheid: een ander gezichtspunt. Toekomstperspectief vanuit vier disciplines*, Brugge: Die Keure, 2000, p. 189–241.

LECLERCQ, J.-F., Rapport introductif, in: FAGNART, J.-L. (ed.), *1903–2003. Accidents du travail: cent ans d'indemnisation*, Bruxelles: Bruylant, 2003, p. 1–13.

LENAERTS, H., *Inleiding tot het social recht*, Diegem: Kluwer, 1995, 662 p.

MAGNUS, U. (ed.), *The Impact of Social Security Law on Tort Law*, Wien-New York: Springer, 2003.

MAGNUS, U., Compensation for Personal Injuries in a Comparative Perspective, *Washburn Law Journal*, vol. 39, 2000, p. 347–362.

MARKESINIS, B.S., DEAKIN, S. & JOHNSTON, A., *Markesinis and Deakin's Tort Law*, Oxford: Clarendon Press, 2003, 871 p.

MESTREIT, X., Discours de rentrée, *Jurisprudence Liège*, 1905, p. 297 *et seq.*

MONDELAERS, Y., *Het ontstaan en de evolutie van de arbeidsongevallenverzekering in België (1880–1930)*, 2000, unpublished, 118 p.

PEARSON, *Report of the Royal Commission on Civil Liability and Compensation for Personal Injury*, vol. 1, Cmnd 7054-I, London: Her Majesty's Stationery Office, 1978, 545 p.

RAUWS, W., Financiering van schade veroorzaakt door arbeidsongevallen en (nieuwe) beroepsziekten: België als wenkend voorbeeld?, in: FAURE, M. & HARTLIEF, T., *Schade door arbeidsongevallen en nieuwe beroepsziekten*, Den Haag: Boom Juridische Uitgevers, 2001, p. 109–129.

RITTER, G.A., *Social Welfare in Germany and Britain: Origins and Development*, Leamington Spa: Berg, 1983, 211 p.

SAINCTELETTE, CH., *De la responsabilité et de la garantie (accidents de transport et de travail)*, Bruxelles: Bruylant-Christophe, 1884, 258 p.

SCHWITTERS, R.J.S., *De risico's van de arbeid. Het ontstaan van de Ongevallenwet 1901 in sociologisch perspectief*, Groningen: Wolters-Noordhoff, 1991, 337 p.

SIJSES, G.P., *Wet op de arbeidsongeschiktheidsverzekering*, Deventer: Kluwer, 1965, 130 p.

SMULDERS, P.G.W., GRUNDEMANN, R.W.M. & WILLEMS, J.H.B.M., De omvang van het 'risque professionel' in Nederland, *SMA*, 1993, 400–409.

SWAAN, A. DE, *Zorg en de staat. Welzijn, onderwijs en gezondheidszorg in Europa en de Verenigde Staten in de nieuwe tijd*, Amsterdam: Bert Bakker, 1996, 342 p.

TANGHE, F., Het arbeidsongeval: aansprakelijkheid als sociale constructie, *Recht der Werkelijkheid*, 1989, 2, p. 61–75.

VAN DER VORST, P., Esquisse d'une théorie générale du « risque professionnel » et du « risque juridique », *Journal des Tribunaux*, 1975, n° 4916, p. 373–377 et n° 4917, p. 389–394.

VAN LANGENDONCK, J., Arbeidsongeval en aansprakelijkheid, *Tijdschrift voor Sociaal Recht*, 1988, 73–88.

VAN STEENBERGE, J., De Arbeidsongevallenwet: tussen de 19e en de 21ste eeuw, *Belgisch Tijdschrift voor Sociale Zekerheid*, jg. 23, 1990, n° 6–7–8, p. 363–369.

VAN STEENBERGE, J., *Schade aan de mens. Deel I. Evaluatie van de arbeidsongeschiktheid in het recht*, Berchem/Antwerpen – Amsterdam: Maarten Kluwers Internationale Uitgeversonderneming, 1975.

VELDKAMP, G., Enkele critische opmerkingen over het risque professionnel, *Sociaal Maandblad*, 1945, p. 125–134.

VELDKAMP, G., *Individualistische trekken in de Nederlandse sociale arbeidsverzekering: een critisch onderzoek naar de grondslagen der sociale arbeidsverzekering*, Alphen aan den Rijn: Samsom, 1949.

VERVLIET, V., *De buitencontractuele aansprakelijkheid bij professionele risico's: een onderzoek naar de betekenis en draagwijdte van de buitencontractuele aansprakelijkheid doorheen de evolutie van de professionele risicoverzekering*, Antwerpen: unpublished, 2006, 666 p.

WILSON, A. & LEVY, H., *Workmen's Compensation*, vol. 1: Social and Political Development, London: Oxford University Press, 1939, 328 p.

WICKENHAGEN, E., *Geschichte der gewerblichen Unfallversicherung*, 1980.
WIKELEY, N.J., *Compensation for Industrial Disease*, Aldershot: Dartmouth Publishing, 1993, 221 p.
WIKELEY, N.J., OGUS, A.I. & BARENDT, E.M., *The Law of Social Security*, 5th ed., London: Butterworths LexisNexis, 2002, 803 p.

Industrial Accidents and Occupational Diseases: Some Empirical Findings for The Netherlands, Belgium, Germany and Great Britain

*N.J. Philipsen**

1. Introduction

This paper, which has been composed within the framework of the 'Shifts in   1
Governance' project, contains an impression of first empirical data that was
collected on industrial accidents and occupational diseases in four European
countries: The Netherlands, Belgium, Germany and Great Britain. The analysis so far is modest: We just looked for data on the incidence and scope of
work injury in those countries. In a next phase these data could be used to
analyse the effectiveness of various compensation systems, such as tort law
and private insurance on the one hand and no-fault compensation funds and
social security on the other. This paper addresses seemingly simple questions,
such as: what exactly are the sources of compensation for work injury in each
of those countries? And how much do these sources of compensation pay out
in cases of industrial accidents or occupational diseases? It will be very difficult, however, to answer these questions, as in all four countries under review
here the compensation system consists of several different layers.

In addition to the *compensation* aspect, the *prevention* aspect of the various   2
compensation systems needs to be considered. In that respect the central question is whether the different systems have a deterrent effect on the number of
industrial accidents and occupational diseases. This question is particularly interesting when tort law is involved, since it has often been claimed in the literature that (in general) the credible threat of a liability suit has a deterrent effect

* Maastricht University, METRO Institute for Transnational Legal Research, PO Box 616, 6200
  MD Maastricht, The Netherlands, E-Mail: niels.philipsen@facburfdr.unimaas.nl. The author
  would like to thank the participants in the Shifts in Governance pre-meeting in Vienna (7–8
  October 2005) for valuable comments and questions.

on the actual number of accidents.[1] However, to measure the effect on prevention, more and other data are needed. That will not be the central focus of this paper.[2]

3   In order to analyse the effectiveness of any system as far as compensation and prevention is concerned, one could either compare countries with different systems or one could try to identify shifts in the system of one country. In the 'Shifts in Governance' project we focus on the latter, i.e.: Can we identify shifts from one compensation system to another, or shifts within one compensation system, in any of the countries under review?[3] And if so, can we find some empirical data on the effects of these shifts?

4   However, first some limitations of the approach taken in this paper need to be clarified. Ideally, one would like to present statistical data concerning, *inter alia*, the following indicators:

- (changes in) the number of work-related incidents, relative to the number of people employed
- (changes in) the number of claims and settlements
  - as a result of an industrial accident
  - as a result of an occupational disease
- (changes in) the total amounts of compensation paid to the victim-employees
- (changes in) the composition of the compensation paid: Who pays the various components of the damage?

Unfortunately, these data are often not available as such, because there has been little or no registration of such indicators. Therefore, one has to look for different – more general – indicators first and try to work from there. This paper comprises some first steps in trying to find answers to the questions mentioned above.

5   The paper deals with each of the four countries separately. First the Netherlands will be considered, then Belgium, Germany and finally Great Britain. The structure of the analysis will be similar for each country. After a short overview of the compensation system and the sources of compensation in the country concerned, we will present some data on the (estimated) number of industrial accidents and occupational diseases in the particular country. When

---

[1] For an analysis of the deterrence perspective on tort law, see, *e.g.*, S. Shavell, Liability for Harm versus Regulation of Safety, [1984] *Journal of Legal Studies* (JLS) 13, 357–374 and D. Dewees/ D. Duff/M. Trebilcock, *Exploring the Domain of Accident Law: Taking the Facts Seriously* (1996). See also my other contribution to this book (which includes additional references) and the contribution to book 3 by Ogus.

[2] This could be the subject of further research.

[3] This does not mean that we ignore the comparative analysis. It only means we will focus primarily on the analysis of shifts within countries. For a definition of shifts see also the introduction to this book by Hartlief and Klosse.

addressing compensation and prevention of occupational risks in a particular country one should at least have some information about the actual number of incidents and, if possible, developments over time which are related to that number. One might even attempt to relate such data to goals of prevention, although one has to be very careful in doing so.[4] After all, there are many factors influencing the actual number of accidents in addition to regulation of safety and the threat of a liability suit, such as the proportion of the labour force employed in high risk industries, the proportion of flexible workers, work pressure, the general attitude towards risk, etc. Next, we will turn to data on sources of compensation. The questions one would theoretically like to answer are clear. How much does social security pay out exactly via the different sources of compensation? And if tort law applies: What are the average amounts of settlements and what are the average damage payments if it comes to a lawsuit? As already stated above, in practice it is very difficult to find concrete answers to these questions. Therefore, first some general data on social security systems and – where possible – tort law and insurance will be presented, insofar as they apply to our topic of industrial accidents and occupational diseases. Further research should of course elaborate on these points.

This paper is set up as follows. After this introduction the situation in the Netherlands (section 2), Belgium (3), Germany (4) and Great Britain (5) is addressed. Finally, section 6 of this paper presents a few concluding remarks.

2. The Netherlands

2.1 Overview[5]

Compensation for work-related harm in the Netherlands is not based on the idea of *risque professionnel*. Rather this type of harm is considered as a *risque social*.[6] This has been the case since 1967, when the *Wet op de Arbeidsonge-schiktheidsverzekering* (Disability Insurance Act, WAO) was introduced. Until 1967 the *Ongevallenwet* (Industrial Injuries Act) applied, which was based on compulsory private insurance for employers. Hence we can clearly speak of a 'shift in governance' here.

Let us turn to the situation in the Netherlands as it is today. In cases of work-related harm, a victim might obtain compensation via different 'compensation layers': social security, private insurance and (sometimes) tort law. According to labour law (art. 7:658 of the Civil Code), employers are under certain conditions liable for any work-related harm. Moreover, article 7:629 of the Civil Code contains an obligation for employers to keep paying wages (at

---

[4] As mentioned above, that may be the subject of a subsequent paper.
[5] For a more elaborate overview I refer to the contributions to this book by Engelhard and Hoop and to A. van de Goor, *Effects of Regulation on Disability Duration* (1997), 11–48.
[6] See also S. Klosse, Schadevergoeding via Sociale Zekerheid en Aansprakelijkheidsrecht: Communicerende Vaten?, in: M. Faure/T. Hartlief (eds.), *Schade door Arbeidsongevallen en Nieuwe Beroepsziekten* (2001), 1–18.

least 70%) during the employee's first two years of absence from work, if this absence results from illness. As a result, employers take out insurance to cover these risks. Up to 1996, the *Ziektewet* of 1929 (Sickness Benefits Act) provided an important part of the short-term compensation of lost income. However, since the privatisation of this Act in 1996, its relevance has been limited to special categories of employees, *e.g.* employees who do not have a permanent contract. Long-term income losses, i.e. after the first two years of absence from work, were until 2006 covered by the WAO, already mentioned above.[7] As far as medical costs are concerned, until 2006 the *Ziekenfondswet* of 1967 (Health Insurance Act) applied for the lower incomes, whereas higher incomes were covered by private health insurance.[8] Additionally, the *Algemene Wet Bijzondere Ziektekosten* of 1967 (Exceptional Medical Expenses Act) applies for all Dutch citizens. Finally, victims of work injury might try the path of tort law in order to receive additional compensation, notably for the top of the income loss, property damage and pain and suffering. Naturally, if tort cases cannot be solved by settlements between employees and employers/insurers, they will eventually lead to lawsuits.

7   Before addressing the statistics on industrial accidents and occupational diseases, it is interesting to point out some (minor) shifts that took place in the Netherlands in the early and mid-1990s. In 1992 the new Civil Code, which included various new grounds of strict liability, was introduced in the Netherlands, whereas in 1993 several changes[9] took place with respect to the WAO. This also led (eventually) to the privatisation of the Sickness Benefits Act, already mentioned above, as well as to the introduction of the so-called *Wet Pemba*, which links the social premiums paid by employers for the WAO system more specifically to characteristics of the particular employer. Another minor shift occurred in 2002, when the *Wet Verbetering Poortwachter* (Gatekeeper Improvement Act) was introduced. This Act imposes some concrete measures on employers and employees to promote the reintegration of sick employees.

2.2   Industrial Accidents

8   We noted in the previous section that the Netherlands does not have a specific (mandatory) insurance against occupational risks. This explains why registration of industrial accidents has until recently been very poor, as will be shown

---

[7] For obvious reasons the WIA (which replaced the WAO in 2006) will not be discussed here. There are, after all, no empirical data yet on its effects. For a discussion of the WIA see the contributions to this book by Engelhard and Hoop.

[8] Again, there have been some changes since January 2006, which will not be discussed here (cf. fn. 7).

[9] A. van de Goor (*supra* fn. 5), 39–42, lists these changes. They are: the introduction of a new disability criterion (the illness or injury that causes work incapacity has to be assessed with 'medical objectivity' and the kind of labour that is considered when determining residual earning capacity has changed); the division of benefits into two stages (benefits related to last earned wage and continuation benefits); and the introduction of a number of new conditions for benefit entitlement in the WAO (including application requirements and a training obligation).

in more detail below. While in *risque professionnel* systems industrial accidents and occupational diseases are registered for the purpose of compensation, the purpose of registration in the Netherlands is prevention (via provision of information). In that respect the *Arbeidsomstandighedenwet* (Working Conditions Act) contains some provisions directed at the prevention of industrial accidents and the reporting of such accidents. This Working Conditions Act was introduced in 1980 to replace the *Veiligheidswet* of 1895 (Safety Act). However, despite the obligation of employers to report all serious accidents to the Labour Inspectorate, registration of accidents in the Netherlands has long been unreliable. According to Popma and Venema (2006) this is mainly attributable to the fact that in the Netherlands[10]

> 'no notification of an occupational accident to an insurance body is required. Insurance based systems […] have reporting procedures mainly based on the notification of the accidents to the insurance company, which spurs companies to report [NP: the authors refer to Eurostat (2001)]. The legal obligation to report all 'serious' accidents to the Labour Inspectorate […] is often being evaded. […] Finally, the data collected on occupational accidents by other instances differ in the way they collect their data, the target population, the definitions and classifications used'.

In 2001, the Dutch Ministry of Social Affairs and Employment asked research institutes *TNO Arbeid* (TNO Work and Employment) and *Stichting Consument en Veiligheid* (Consumer and Safety Foundation) to bridge the above-mentioned methodological gaps. After two years of test runs, the first so-called *Monitor Arbeidsongevallen* (Industrial Accidents Monitor) was published in 2004.[11] This report combines data from several sources[12] and defines three types of industrial accidents: lost time accidents, serious accidents and fatal accidents. Lost time accidents (*arbeidsongevallen met letsel en verzuim*) are those resulting in absence from work for at least one working day. The report estimates that about 103,000 lost time accidents occurred in 2002. The number of serious accidents in 2002, i.e. accidents leading to hospitalisation, is estimated at 3,500. The number of fatal accidents (resulting in death) mentioned in the report is 91.[13] All numbers are inclusive of work-related road traffic accidents and exclusive of commuting accidents. Data about the distribution of accidents by age, sex, origin and other personal characteristics, as well as the distribution of accidents by industry sector, are available as well.

---

[10] J.R. Popma/A. Venema, Occupational Accidents in the Netherlands, in: Smulders (ed.), *Worklife in the Netherlands* (2006), 169.

[11] A. Venema/A. Bloemhoff, *Monitor Arbeidsongevallen in Nederland 2002* (TNO Arbeid 2004).

[12] The Industrial Accidents Monitor combines input from Netherlands Statistics (accidents resulting in absence from work for at least one day and fatal accidents), the *Letsel Informatie Systeem* developed by the Consumer and Safety Foundation (accidents leading to hospitalisation) and the Labour Inspectorate (fatal accidents).

[13] The same numbers can be found in SZW, *Arbobalans 2004* (2004), 28. Numbers provided in previous editions of this *Arbobalans*, however, are based on different sources and cannot be compared to the 2002 data.

9   Because reliable pre-2002 data are unavailable, it is impossible to analyse
     trends in the number of industrial accidents. Apart from numbers for the year
     2002, the Industrial Accidents Monitor only provides data for 2000 and 2001.
     These data are presented in Table 2.1 below. This table also includes the num-
     ber of accidents per 100,000 persons employed[14], as well as numbers for the
     year 2003.[15]

Table 2.1 Number of Industrial Accidents and Accident Rates (2000–2003)

		Fatalities	Serious accidents	Lost time
2000	Number	119	4,100	103,000
	Per 100,000	1.5	53	1,300
2001	Number	115	3,500	95,000
	Per 100,000	1.5	45	1,200
2002	Number	91	3,500	103,000
	Per 100,000	1.2	45	1,300
2003	Number	104	3,200	93,000
	Per 100,000	1.3	40	1,200

10   TNO Work and Employment has been developing an instrument to assess the
     direct and indirect costs of work-related injuries to the *companies* that employ
     or employed the victims. Rough estimates range from € 250 for one day lost
     accidents to € 430,000 for serious accidents with lasting incapacitation and a
     successful tort claim.[16] A distinction is made between various categories of
     costs for the employer, including loss of production, legal sanctions (such as
     fines) and liability costs (claims from employee or insurer). These cost catego-
     ries are discussed briefly in the Industrial Accidents Monitor. Popma and Ve-
     nema (2006) mention that also estimates about the costs of public healthcare
     and income loss for the *employee* are hampered by shaky data, however:[17]

     'The Consumer and Safety Foundation is presently elaborating indica-
     tors. A rough estimate of the direct costs of work related first aid at hos-
     pitals was € 42 million in 2001. The proportion of disablement due to ac-
     cidents at work may be calculated off the cuff, using statistics from
     claims made to the WAO. Of all the disabled employees, 5.4% have the
     diagnosis 'injury' [source: UWV (2004)]. Some 10% of the most serious
     injuries being work related [source: RIVM (2004)], one may conclude
     that 0.5% of all WAO benefits may be attributed to accidents at the work-
     place – hence, over € 50 million [source: UWV (2004)]'.

---

[14] A. Venema/A. Bloemhoff (*supra* fn. 11), 25.
[15] A. Venema/A. Bloemhoff, *Monitor Occupational Accidents in the Netherlands 2003: Summary
      and tables* (TNO Quality of Life 2005).
[16] A. Venema/A. Bloemhoff (*supra* fn. 11), 27–28.
[17] J.R. Popma/A. Venema (*supra* fn. 10), 171.

Unfortunately, these are apparently the rough kinds of computations one has to make in order to filter relevant data from aggregated statistics, such as (in this case) the total amount of disability benefits received. We will come back to this problem below.

2.3   Occupational Diseases

Arbodiensten (Occupational Health and Safety Services)[18] have since 1999 been obliged to notify occupational diseases to the *Nederlands Centrum voor Beroepsziekten* (Netherlands Center for Occupational Diseases, NCvB).[19] Indeed, much information can be found on the website of NCvB. However, like in most European countries the level of underreporting is considerable.[20] The official numbers are given in Table 2.2:[21]

11

Table 2.2 Number of Occupational Diseases (2000–2004)

	2000	2001	2002	2003	2004
Binnengekomen meldingen Number of notifications	n.a.	6,384	6,639	7,147	n.a.
Geaccepteerde meldingen Accepted notifications[22]	6,063	5,593	5,335	5,973	5,778

The increase in the number of notifications in 2003 was most probably not due to an actual increase in the number of occupational diseases but resulted from a higher notification percentage. Figures about the distribution of occupational diseases by sector, diagnosis, age, sex and even the number of notifications per Occupational Health and Safety Service are also available from NCvB. There are no data on the average costs per incident. Here we will only provide a top 10 list of accepted notifications in 2003 by disorder:[23]

---

[18] Occupational Health and Safety Services are agencies responsible for the enforcement of the Working Conditions Act. They, *inter alia*, advise on working conditions, monitor and guide disabled employees, etc.

[19] NCvB (*Nederlands Centrum voor Beroepsziekten*) is a research institute linked to the University of Amsterdam. Website: http://www.beroepsziekten.nl.

[20] Nederlands Centrum voor Beroepsziekten, *Signaleringsrapport Beroepsziekten* (2004), 9. For the reasons of underreporting see p.15 of this report and Eurogip, *Survey on Under-Reporting of Occupational Diseases in Europe* (2002). The Occupational Diseases Bureau (BBZ, see below for more information) claims on its website that 25,000 employees contract an occupational disease each year in the Netherlands, which would imply that the level of under-reporting is huge indeed: see http://www.bbzfnv.nl (in Dutch only).

[21] Source: NCvB website. The same numbers can be found in: Ministerie van Sociale Zaken en Werkgelegenheid, *Arbobalans 2004* (2004), 25.

[22] Some reportings are not accepted because they do not fulfil the criteria that define an occupational disease.

[23] NCvB, *Annual Report* (2003), 6.

Table 2.3 Occupational Diseases by Disorder (2003)

	Number of notifications	%
1. Occupational hearing loss	1506	25.2
2. Adjustment disorders	830	13.9
3. RSI of shoulder/upper arm	687	11.5
4. Lateral epicondylitis	303	5.2
5. Burn out	295	4.9
6. RSI of elbow/lower arm	266	4.5
7. RSI of wrist and hand	243	4.1
8. Contact dermatitis	202	3.4
9. Other disorders of soft tissues due to pressure and (over)strain	166	2.8
10. Carpal tunnel syndrome	79	1.3
Total	4577	76.6
Remaining disorders	1396	23.4
Total	**5973**	**100**

12   It is difficult to find reliable pre-2000 data on occupational diseases, as there was simply no good registration until then. Only some very rough estimations of incidents and costs can be found in the literature. Faure, Hartlief and Philipsen (2006) refer to research carried out in 1995 at the request of the Netherlands Association of Insurers, in order to examine the potential volume of cases dealing with industrial accidents and occupational diseases. At that time:[24]

> 'predictions were made that the number of accidents at work would stabilize, or at least not increase [NP: the authors refer to Haazen/Spier (1996)], but the reverse appeared to be true for occupational diseases. Van Mierlo[25] outlined various scenarios for the potential growth of the number of occupational diseases in the Netherlands whereby even in a minimum scenario it was estimated that the absolute costs of occupational diseases would increase from NLG 58 million (€ 26.3 million) in 1990 up to NLG 259 million (€ 117.5) in 2005. This fear was mainly based on the expected increase in the number of asbestos-related illnesses, but also concerned new unknown occupational hazards like RSI (Repetitive Strain Injury) and OPS (Organic Psychic Syndrome)'.

13   Worth mentioning here, finally, are the activities of the *Bureau Beroepsziekten FNV* (Occupational Diseases Bureau, BBZ). The BBZ is a foundation of the largest Dutch labour union FNV, which works on a 'no win no fee' basis and which provides legal assistance to (supposed) victims of occupational diseas-

---

[24] M.G. Faure/T. Hartlief/N.J. Philipsen, [2006] Funding of Personal Injury Litigation and Claims Culture: Evidence from the Netherlands, *Utrecht Law Review*, Vol. 2, 8.
[25] J.G.A. van Mierlo, Economische Inschatting van de Evolutie in de Kosten, in: M.G. Faure/T. Hartlief (eds.), *Verzekering en de Groeiende Aansprakelijkheidslast* (1995), 236.

es. This legal assistance includes everything from retrieving medical and other information from employers and Occupational Health and Safety Services, filing (where possible) tort claims against a liable employer and negotiating with insurance companies about the amounts of compensation, to – when it comes to that – legal procedures. In the period May 2000 – September 2005, the BBZ had already received more than 3,000 notifications from employees, of which 900 were taken up. In 700 of these cases the BBZ judged that there were grounds to hold the employer liable. Although the BBZ dossiers are confidential, they might contain some interesting information, *e.g.* about settlements, compensation paid, reactions by employers, etc.

2.4 Amounts of Compensation Received

While, ideally (as noted in Section 1) one would like to show statistical data on the *amounts* of compensation received in case of an industrial accident or occupational disease, as well as data on damage payments paid out (relative to the total harm) when a case comes to court, such data are very difficult to obtain for the Netherlands. Even seemingly more simple-to-obtain statistics, such as the estimated *number* of tort cases and settlements, appear to be very hard to find. Explanations for this are obvious. First: registration in the Netherlands has always been pretty bad, at least until very recently. As discussed above, this follows from the fact that compensation for work-related harm in the Netherlands is considered as a *risque social*. As a result, incentives to collect specific data on occupational risks were lacking for a long time. Second, insurers and personal injury lawyers are often not very co-operative in the provision of information even if they have it. To some extent this may be a result of competition in the respective markets, i.e. the fear of insurers and personal injury lawyers to play into the hands of competitors by bringing confidential information into the open.[26]

2.4.1 Social Security

In the previous sections we presented some data on the total number of work-related incidents and some rough indications of corresponding costs. Here a link will be made to aspects of social security. In that respect *Statistics Netherlands (Centraal Bureau voor de Statistiek*, CBS) provides some general statistics concerning indicators such as the number of recipients of various kinds of benefits and the total (i.e. aggregated) amounts paid out by the government. These numbers, however, do not tell us whether the benefits paid out were related to industrial accidents and occupational diseases or to other causes. Therefore, these numbers as such do not provide an answer to the problem statements defined within the context of the 'Shifts in governance' project. As the next step, we need to think about how we could link these (aggregated) statistics to the area of occupational risks.[27] More specifically: How can we refine these data in such a way that they give information on the actual compen-

---

[26] See also M.G. Faure/R. Van den Bergh, Restrictions of Competition on Insurance Markets and the Applicability of EC Antitrust Law, [1995] *KYKLOS* 48, 65–85, for an analysis of insurers' behaviour.

[27] See for an example my quote from Popma and Venema (*supra*, fn. 17).

sation for industrial accidents and occupational diseases? As various compensation sources are involved, such as the WAO, Sickness Benefits Act, Health Insurance Act, Exceptional Medical Expenses Act and compensation received from the employer/insurer, this is a challenging task.

Table 2.4 shows some figures on the number of benefit recipients in the Netherlands in the period 1986–2004. Longer time series are available: Sickness benefits data are available as from 1930 and disability benefits as from 1901. Most noticeable in the table is a large drop in the number of sickness benefits after 1995. This probably relates to the privatisation of the Sickness Benefits Act.[28] Furthermore, we see that the number of recipients of disability benefits has been increasing most of the time in the period under review, except for the mid 1990s. Of course, we have to keep in mind that the number of WAO benefits is also related to the size of the labour force (i.e. the number of employed and unemployed people).

Table 2.4 Benefit Recipients (1986–2004)

Number of benefit recipients (*uitkeringsontvangers*) × 1,000[29]			
Year	Sickness benefits *Ziektewet*	Disability benefits *Arbeidsongeschiktheidsuitkeringen:* WAO/Wajong/WAZ	WAO
1986	236	783	–
1987	252	793	–
1988	261	814	–
1989	262	845	–
1990	291	879	–
1991	294	904	–
1992	289	913	–
1993	292	921	–
1994	172	894	–
1995	169	872	–
1996	41	873	–
1997	92	889	–
1998	103	907	729
1999	107	926	746
2000	106	957	772
2001	106	981	794
2002	79	993	802
2003	66	982	787
2004	n.a.	961	763

[28] Discussed in Section 2.1.
[29] Source: Statistics Netherlands (CBS), 2004–2005. As regards the number of occupational disability benefits, the same numbers are mentioned by UWV (Uitvoering Werknemersverzekeringen).

In addition to statistics on the number of benefit recipients, the CBS has col- 16
lected some statistics relating to sickness absence and benefits paid. As regards disability benefits, Table 2.5 gives data for the more recent years. Again,
these data are aggregated: It concerns total amounts paid by the Dutch government. Longer time series for the period 1901–2003 are available. We see a
large drop in 1993, probably caused by the changes in the WAO in 1993.[30]

Table 2.5 Disability Benefits: Amounts Paid by the Government (1985–2004)

Year	Amounts paid (*uitgekeerde bedragen*) × mln. Euro[31]
	Disability benefits
	Arbeidsongeschiktheidsuitkeringen: WAO/Wajong/WAZ
1985	10,023
1986	10,410
1987	10,725
1988	11,146
1989	11,730
1990	14,042
1991	14,500
1992	14,929
1993	8,389
1994	8,258
1995	7,883
1996	7,738
1997	8,483
1998	9,227
1999	9,819
2000	10,232
2001	10,742
2002	11,454
2003	11,459
2004	11,454

2.4.2 Torts

Data on the estimated number of claims leading to settlements and lawsuits 17
and the money amounts involved in these claims are very difficult to find. As
expected, the CBS appears not to have any information about the number of
cases and settlements resulting from industrial accidents and occupational dis-

---

[30] See Section 2.1 and A. van de Goor (*supra* fn. 5), 42–44.
[31] Source: Statistics Netherlands (CBS), 2004–2005.

eases. Possible other sources are insurers (organised in the Netherlands Association of Insurers, *Verbond van Verzekeraars*)[32] and personal injury lawyers,
especially those dealing specifically with labour cases such as Arboclaim.
However, as mentioned above Dutch insurers and lawyers are usually not very
co-operative for reasons of confidentiality. Furthermore, the dossiers of the
Occupational Diseases Bureau (BBZ), mentioned above, might be interesting
in this respect.

18   Some general information can be found in the annual statistical report of the
     Netherlands Association of Insurers, which includes a table with key figures
     on general liability. These figures are obviously not restricted to employers' li-
     ability for work-related injuries. Table 2.6 below provides an extract from this
     table.[33]

Table 2.6 Key Figures Insurance: General Liability (1999–2004)

	1999	2000	2001	2002	2003	2004
Claims						
Gross incurred damage	404	415	470	513	563	603
Of which private	141	136	150	160	163	197
Of which companies	262	277	320	353	400	406
Average amount per claim (€)						
Private	n.a.	390	444	479	509	600
Enterprises	n.a.	3,902	4,595	5,369	6,272	6,768
Claim frequency: Number of claims per 100 policies						
Private	n.a.	5.3	5.0	5.0	4.6	5.1
Enterprises	n.a.	11.1	12.1	11.5	10.4	10.0

The table shows that the gross incurred damage by insurers has been increasing in recent years, while the claim frequency has been decreasing. The
explanation for this apparent contradiction is the average amount per claim,
which had increased sharply between 2000 and 2004 (especially with respect
to enterprises). Of course, one has to keep in mind that these numbers concern
general liability. Unfortunately, we cannot filter the data that specifically considers industrial accidents and occupational diseases.

19   As an indication of the problems involved in empirical research on torts, an in-
     teresting reference is Weterings (1999)[34], who attempted to estimate the aver-
     age number of personal injury claims per year, using a step-by-step approach
     involving three stages: (1) stock-taking exercise of the total number of acci-
     dents leading to injuries; (2) analysis of the number of potential insurance

[32] Website: http://www.verzekeraars.org.
[33] Source: Verbond van Verzekeraars, *Verzekerd van Cijfers* (2005), 24–25.
[34] W.C.T. Weterings, *Vergoeding van Letselschade en Transactiekosten: Een Kwalitatieve en Kwantitatieve Analyse* (1999).

claims resulting from this; and (3) assessment of the actual number of claims filed with the injurer or his insurer. Weterings argues that there is a huge difference between the potential number of claims filed and the actual number. The gigantic detour used by Weterings to get to the estimated number of insurance claims shows us that data are not readily available. With respect to industrial accidents, the figures mentioned in his book, which are all estimations for the early 1990s, are as follows.

Table 2.7 Industrial Accidents: Number of Potential Claims and Claims Filed[35]

Estimated number of victims of industrial accidents per year	
Victims (total)	±300,000–350,000
Medical treatment	±200,000–230,000
Treatment in an outpatients' department/ Specialistic treatment	±120,000
Hospitalisation	±6,000
Permanent injury	±21,000
Deceased victims	±45

Number of victims	Potential claim percentage	Number of potential claims
300,000–350,000	±80–90%	240,000–315,000

Number of victims	Percentage of claims filed	Number of claims filed
300,000–350,000	±10–15%	30,000–52,500

3. *Belgium*

3.1 Overview[36]

In Belgium, occupational risks are considered as a *risque professionnel*. There is a no-fault insurance system, which is an integral part of the social security system and which is paid for by employers collectively. Work injury victims receive a standardised compensation of lost income – whereby a statutorily-determined ceiling (of currently € 32,106 a year for industrial accidents) is taken into account – as well as compensation for healthcare costs. Non-pecuniary losses and property damage are not compensated via this scheme. Civil liability of employers is generally excluded, except for, *inter alia*, intentional wrongs and property damage.

With respect to industrial accidents resulting in personal injury, the *Arbeidsongevallenwet* of 1971 (Industrial Injuries Insurance Act) applies. Victims of such accidents are compensated by their employers, who have to take out private insurance to cover this risk. In the case where an employer is not

---

[35] Weterings (*supra* fn. 34), 15 *et seq.*

[36] For a more elaborate overview I again refer to the contributions by Engelhard and Hoop. See also W. Rauws, Financiering van Schade Veroorzaakt door Arbeidsongevallen en (Nieuwe) Beroepsziekten: België als Wenkend Voorbeeld?, in: M. Faure/T. Hartlief (eds.), *Schade door Arbeidsongevallen en Nieuwe Beroepsziekten* (2001), 109–129.

insured, the *Fonds voor Arbeidsongevallen* (Industrial Injuries Fund, FAO), established by Royal Decree no 66 in 1967, steps in. Before 1971 the Belgian system was based on civil liability of the employer, the risk of which was also born by employers collectively by means of private insurance. However, at the time such insurance was not mandatory.

Compensation for (recognised) occupational diseases is arranged differently: In these cases compensation is received directly from the *Fonds voor Beroepsziekten* (Occupational Diseases Fund, FBZ), which acts as insurer for the private sector and as reinsurer for the public sector. The amount of compensation paid out by this fund is dependent on the disability percentage of the employee and the income lost. With respect to the private sector, legislation consists of the so-called *wetten betreffende de schadeloosstelling voor beroepsziekten* of 1970 (co-ordinated acts on compensation for occupational diseases) and several implementing orders. Different legislation – but with similar content – applies in the public sector.

3.2 Industrial Accidents

21 Unsurprisingly, registration of the number of industrial accidents in Belgium has always been much better than the registration in the Netherlands. According to the Industrial Injuries Insurance Act, Belgian employers are obliged to report all accidents to their insurer within 10 working days.[37] The FAO keeps a register of all such accidents in the private sector.[38] The data are provided to the FAO by insurers. As far as recent numbers are concerned, in 2004 the FAO registered 198,861 industrial accidents, including 21,370 on the road to and from work, 195 fatal accidents and 13,760 accidents leading to permanent disability. The number of private, still active, insurers has over the years decreased to 17. Collectively these insurers manage about 450,000 insurance policies and collect about € 918 million worth of premiums.[39]

In its extensive General Report 2004, the FAO presents interesting data, such as the evolution of the number of industrial accidents in the private sector in the period 1985–2004. Four categories of accidents are distinguished: without consequence (zonder gevolg)[40], temporary disability (*tijdelijke ongeschiktheid*), permanent disability (*blijvende ongeschiktheid*) and fatal (*dodelijk*). A selection is presented in Table 3.1 below.[41]

---

[37] Accidents leading to property damage only are not considered as industrial accidents in the strict sense of the Industrial Injuries Insurance Act; the reporting obligation concerns accidents leading to personal injury.

[38] As of 2000 the FAO also registers public sector accidents. However, this information is somewhat limited and will not be presented in this paper. Detailed figures on public sector accidents can be found on the website of the Federal Public Service of Employment, Labour and Social Dialogue: http://meta.fgov.be/pa/nla_index.htm.

[39] FAO website: http://socialsecurity.fgov.be/faofat, 8 May 2006.

[40] This category includes minor accidents resulting in an absence from work less than one working day. Such accidents are not even registered in many other countries, among them the Netherlands (see previous section). Also Eurostat, which publishes statistics on industrial accidents in EU Member States, does not include accidents without consequence in its figures. The treshold employed by Eurostat is 'three working days lost'.

[41] FAO, *Algemeen Verslag: Dienstjaar 2004* (2005), 112. See also pp. 69–77 of that report.

Table 3.1 Number of Industrial Accidents in the Private Sector (1985–2004)

Year	Temporary disability	Permanent disability	Fatal	Total (including accidents without consequence)
1985	158,994	10,814	206	243,805
1986	154,756	11,944	178	239,412
1987	153,027	11,347	182	237,869
1988	163,595	10,647	169	249,247
1989	182,621	12,064	200	265,930
1990	190,318	12,195	184	276,281
1991	185,538	12,182	184	267,271
1992	173,981	12,133	156	250,959
1993	145,845	12,023	158	213,865
1994	138,913	12,518	152	206,518
1995	109,065	11,586	139	207,869
1996	101,216	11,177	119	196,637
1997	97,574	12,712	130	197,520
1998	103,262	12,258	138	202,274
1999	102,345	12,479	118	199,715
2000	108,409	13,128	139	209,508
2001	110,294	13,742	127	203,171
2002	96,385	11,710	121	184,252
2003	85,823	12,629	100	170,853
2004	82,559	11,751	122	165,472

Number of industrial accidents accepted by insurers (excl. road to and from work)[42]

Unfortunately, the number of accidents per 100,000 persons employed (the indicator used for the Netherlands in Section 2.2 above) is not available as such for Belgium. We might try to compute it ourselves, using data on the working population. However, we would have to be very careful, as the numbers presented above concern private sector accidents only, meaning we would have to consider the working population in the private sector only. Another way of obtaining accident rates is to simply use statistics provided by the FAO and the Federal Public Service of Employment, Labour and Social Dialogue. Table 3.2 presents the *frequency* of industrial accidents (excluding accidents without consequence), defined by the 'number of accidents per million of hours exposed' as estimated by these two sources. We can see that the frequency of industrial accidents has been decreasing in recent years. 22

[42] Source: FAO (*supra* fn. 41). Numbers of accidents on the road to and from work are also available.

Table 3.2 Frequency of Industrial Accidents in the Private Sector (1997–2002)[43]

	1997	1998	1999	2000	2001	2002
Source: FAO	*31.79*	*33.50*	*27.77*	35.50	35.02	31.23
Source: FPS ELS		35.99	35.82	35.61	32.66	30.15

23   When studying industrial accidents in Belgium, another interesting source of information is Prevent, a multi-disciplinary institute directed at the prevention of occupational risks. On its website Prevent presents itself as follows:

> 'Prevent is een multidisciplinair kennisinstituut gericht op de preventie van beroepsrisico's door de bevordering van de kwaliteit van de arbeidsomstandigheden en de verbetering van de arbeidsorganisatie. Het instituut verleent ondersteuning, advies en informatie, en dit zowel voor bedrijven en instellingen als voor de arbeidsongevallenverzekeraars, de externe diensten voor preventie en bescherming, de beroepsverenigingen, de sociale partners, de overheid en andere maatschappelijke actoren'.[44]

One of the documents published regularly by Prevent is *Statistieken Arbeidsongevallen en Beroepsziekten* (Figures on Industrial Accidents and Occupational Diseases). This document contains information on accident risks per sector, age group, sex, company size, etc., as well as specific details about the types of accidents that occur. Prevent predominantly uses information provided by the FAO and the Federal Public Service of Employment, Labour and Social Dialogue.

3.3   Occupational Diseases

24   As mentioned above, the FBZ takes care of the compensation (i.e. the benefits) of people with an occupational disease. The amount of compensation is dependent on the disability percentage and will always be paid out if the occupational disease in question is on the officially recognised list of occupational diseases. This is called the list system (*lijstsysteem*). There is also an open system (*open systeem*), which applies when occupational diseases are not on this list, albeit under very strict conditions: About 93% of the cases are rejected.[45] Detailed information about the computation of the standardised compensation can be found on the FBZ website.[46]

---

[43] Source: Table from Prevent, *Statistieken Arbeidsongevallen '02 en Beroepsziekten '02'03* (2004), 8. The FAO introduced a new computation method in 2000, which makes it impossible to compare the new estimations with the pre-2000 data. Hence the italics in the table.

[44] Http://www.prevent.be, 8 May 2006. Translation [NP]: Prevent is a multi-disciplinary institute directed at the prevention of occupational risks by promoting quality of working conditions and improving the labour organisation. The institute provides support, advice and information to companies and institutes, insurers, external prevention and protection services, professional bodies, management and trade unions, the government and other social actors.

[45] Eurogip, *Occupational Diseases in 15 European Countries: Data 1990–2000, New Developments 1999–2002* (2002), 15 and Prevent (*supra* fn. 43), 31.

[46] Http://www.fmp-fbz.fgov.be. See also FBZ, *De Beroepsziekteverzekering voor Werknemers in de Privé-Sector* (2004) and Section 3.4.1 below.

With respect to the private sector, the FBZ is an *insurer* of the risk of occupational diseases: It is therefore authorised to investigate applications, to take decisions and to pay out reimbursements. Employers are mandatorily insured with the FBZ. With respect to the public sector, the FBZ is only authorised to act as a *reinsurer* of provincial and local governments concerning the compensation of occupational diseases.[47]

Statistical information is available in the extensive Annual Report 2004 of the FBZ. The information in this report is detailed: It not only includes the number of occupational diseases grouped per type of diagnosis[48], but also per province, per district, per system (list or open system), etc. As with industrial accidents, distinction is made between the private and public sector. Here we focus on the private sector only.[49] In that respect the Annual Report 2004 presents many figures on applications made for compensation in the case of employment disability (per type of diagnosis, age, sex, etc.), applications made in the case of death, decisions taken by the FBZ, and applications for reconsideration (*herzieningsaanvragen*). Table 3.3 presents information on the number of applications made with the FBZ. According to Prevent, these applications can be made either by employees themselves, the doctor in attendance or the health insurer.[50]

25

Table 3.3 Occupational Diseases: Number of Applications Made in the Private Sector (1975–2004)[51]

Number of claims for recognition (*aantal ingediende aanvragen*)			
Year	List system	Open system	Total
1975	9,643	–	9,643
1976	9,382	–	9,382
1977	9,957	–	9,957
1978	10,517	–	10,517
1979	9,864	–	9,864
1980	9,617	–	9,617
1981	9,073	–	9,073
1982	8,065	–	8,065
1983	7,760	–	7,760
1984	7,054	–	7,054
1985	7,097	–	7,097

[47] FBZ, *De beroepsziekteverzekering* (2002), 11.
[48] Seven groups of occupational diseases have been defined, each in turn consisting of many subgroups.
[49] Information on the public sector is available in various FBZ publications and on the FBZ website. Roughly, the number of occupational diseases in the public sector (in terms of applications made) is about a tenth of the number in the private sector.
[50] Prevent (*supra* fn. 43), 29.
[51] Source: FBZ and own calculations.

Table 3.3 (continued)

Number of claims for recognition (*aantal ingediende aanvragen*)			
Year	List system	Open system	Total
1986	6,995	–	6,995
1987	7,415	–	7,415
1988	8,255	–	8,255
1989	8,579	–	8,579
1990	9,476	–	9,476
1991	9,285	29	9,314
1992	9,340	556	9,896
1993	7,805	599	8,404
1994	7,301	562	7,863
1995	6,589	716	7,305
1996	5,886	656	6,542
1997	5,226	849	6,075
1998	5,176	1,055	6,231
1999	4,905	1,030	5,935
2000	4,965	1,610	6,575
2001	5,295	1,503	6,798
2002	5,334	1,174	6,508
2003	5,265	934	6,199
2004	5,570	883	6,453

Although no clear trend can be discerned, it seems the number of applications made with the FBZ has decreased somewhat since the 1970s and 1980s. Moreover, the number of applications via the open system, introduced in 1991, seems to have stabilised already – the most recent years even show a decrease. This might be related to the high rate of rejection.

26 The FBZ does not have figures on the 'number of diseases per millions of hours exposed', since they are simply not available at national level.[52] However, information on the claims for recognition *per 100,000 insured persons* can be found in a report published by the European Forum of Insurances against Accidents at Work and Occupational Diseases. Table 3.4 shows these numbers for the years 1990–2000, along with numbers of *recognised* occupational diseases per 100,000 insured persons.

---

[52] See also Prevent (*supra* fn. 43), 30.

Table 3.4 Occupational Diseases: Numbers per 100,000 Insured Persons (1990–2000)[53]

Year	Number of claims per 100,000 insured persons	Recognised ODs/ notified ODs ratio	Number of recognised ODs per 100,000 insured persons
1990	431	43.2%	186
1991	423	46.8%	198
1992	450	61.4%	277
1993	392	60.6%	237
1994	368	51.4%	189
1995	336	60.9%	204
1996	299	53.4%	160
1997	274	49.5%	136
1998	275	52.0%	143
1999	256	39.0%	100
2000	277	40.5%	112

With respect to the claim rate, the data again show a downward trend. Although this may to some extent be linked to progress made in the area of prevention, the European Forum report presents various other possibilities. It suggests that the downward trend can be explained partly by the fact that Belgian employees do not always apply for recognition for fear of losing their jobs. Also, like in the Netherlands, there is some under-reporting which cannot be measured. More importantly however, the report notes that the evolution of economic activities (closing of mines, incorporation of new diseases such as psychosocial diseases) has not yet been fully taken into account in the system. A final explanation put forward in the report is that:

> 'the recognition system possibly discourages workers from undertaking long, tedious procedures (the examination time is 11 months under the list system and 19 months under the open system). In particular, the introduction of a non-listed system apparently does not make it possible to 'retrieve' cases of diseases which might have a relation with work, because the recognition conditions are very severe (93% of cases are rejected)'.[54]

One can observe from Table 3.4 that the overall ratio of recognised to notified occupational diseases was less than 50% in Belgium in 2000, which is not exceptional. On the contrary: This applies for many European countries, among them Germany (where this ratio is even lower as we will show in Section 3.4).

Table 3.5 presents the diseases giving rise to the greatest number of claims for recognition within the framework of the list system. We only provide a small

27

---

[53] Source: Eurogip (*supra* fn. 45), 31. The data provided in this table concern the private sector only, and connect with the data provided in Table 3.3.
[54] Eurogip (*supra* fn. 45), 15.

selection here; the Prevent and FBZ reports give more in-depth information about the number of claims according to diagnosis or agent and the evolution in this number. The category 'osteoarticular diseases' has been leading the list for years. The category 'asbestos-related diseases', one that has always been high in the list, is missing here because it is split up into different categories by the FBZ. Combined, however, asbestos-related diseases would be found somewhere at the bottom of this top 5 list. Of all the *open system* applications made with the FBZ (missing here), 94% concerns various forms of locomotor apparatus disorders (*aandoeningen van het bewegingsapparaat*) and muscle disorders, such as tendinitis, RSI and back complaints.

Table 3.5 Occupational Diseases According to Diagnosis or Agent in the Private Sector (2003)[55]

Diagnosis or agent	2003
1. Osteoarticular diseases due to mechanical vibrations	2,296
2. Hearing loss and noise-induced deafness	597
3. Neuroparalysis due to pressure	489
4. Skin diseases	315
5. Silicosis	297

3.4 Amounts of Compensation Received

28   As in Section 2.4, we present some first empirical data on the (aggregated) *amounts* of compensation received by employees in cases of work-related injuries. In Belgium this compensation is received mainly from insurers and the two funds, viz. the Industrial Injuries Fund, FAO and the Occupational Diseases Fund, FBZ. Fortunately, both the FAO and FBZ keep a quite detailed registration, so information is relatively easy to obtain compared to the Netherlands. The role of tort law is limited in Belgium insofar as workplace incidents are concerned.

3.4.1 No-Fault Insurance System

29   We mentioned before that in Belgium the no-fault insurance system is an integral part of the social security system. In this section we present some first (aggregated) data on the amounts of compensation paid out via this scheme to victims of work injury. Again, we focus on compensation for *income losses* caused by personal injury and leave out the issue of compensation for healthcare costs, which is a related but slightly different issue.

With respect to industrial accidents the FAO provides information about the computation of the standardised minimum compensation (*minimumvergoeding*)[56] and about the evolution of the various kinds of supplementary payment

---

[55] Sources of Table 3.5 and the accompanying text: Prevent (*supra* fn. 43), 34; Eurogip (*supra* fn. 45), 17; own calculations.

[56] This minimum compensation is calculated on the basis of the following formula, which applies in cases of permanent disability: basic amount × disability percentage × reassessment coefficients × adjustment coefficients (*minimumvergoeding = basisbedrag × percentage × herwaarderingscoëfficienten × vereffeningscoëfficienten*). A similar formula exists for benefits in cases of decease.

(*bijslagen*), the number of people eligible for supplementary payment and the *sociale prestaties*, i.e. the amounts paid out by the FAO itself. This information is detailed, but not very relevant for the purposes of our research. After all, with respect to the actual compensation of victims, the FAO can mainly be considered as a kind of 'safety net': When employers are not insured despite the legal obligation to do so, the FAO steps in. Moreover, the FAO will usually try to take recourse against the employer in such cases (which is difficult, as it often concerns employers who have already left the market).

Although more difficult, it would be much more interesting to present infor- 30 mation on the amounts of compensation paid out by insurers. Prevent provides an estimation of the average costs of industrial accidents, based on the amounts of compensation paid out by insurers. For the year 2002 the average costs are estimated at € 3,399 per accident and at € 5,878 if only accidents causing temporary and permanent disability are considered.[57] According to Prevent these numbers are incomplete, because the 'monetary reserves of insurers could not be taken into account in the computations'. Complete data are only available for the period 1989–1999 and are provided in Table 3.6.[58] This table shows an increase in the average costs of industrial accidents for almost the entire period, which can largely be explained by the increasing costs of serious accidents. In 1999, for the first time, the costs per accident decreased.

Table 3.6 Industrial Accidents: Average Costs 1989–1999 (€)

1989	1990	1991	1992	1993	1994	1995	1996	1997	1998	1999
2,618	2,640	2,739	2,995	3,418	3,694	3,924	4,130	4,279	4,321	3,912

Turning to occupational diseases, Table 3.7 presents some general statistics on 31 amounts of compensation paid out. Presented are the cumulative number of victims of occupational diseases (temporary and permanent disability) and the total compensation paid out by the insurance system, i.e. via the Occupational Diseases Fund, FBZ. The numbers concern the years 1997 and 2003, to allow for some comparison.[59] Included also are the five 'most important recognised occupational diseases'.

It is striking that silicosis, mostly among former mine workers, is still so high on this list – even though the number of victims gradually decreased between 1997 and 2003. Silicosis is still the most expensive occupational disease in Belgium. A gradual decrease in the number of victims can also be noted for asbestosis, at least since 1998 (not included in this table). Osteoarticular diseases and skin diseases show a small increase both in numbers and damage payments. Overall the damage payments were € 336.3 million in 2003, which

---

[57] Prevent (*supra* fn. 43), 26.
[58] Source: Additional information provided by Prevent (conversion from Belgian francs to euros by author).
[59] Numbers for the intervening years have been left out. They can be found in the original sources (*infra* fn. 60).

Table 3.7 Occupational Diseases: Number of Victims with Temporary or Permanent Disability (Cumulative) and Total Amount of Damage Payments (× € Million) in 1997 and 2003[60]

Diagnosis	1997		2003	
	Number	Damage	Number	Damage
Osteoarticular diseases due to mechanical vibrations	31,468	85.3	32,794	85.1
Silicosis	27,798	168.6	16,417	119.3
Hearing loss and noise-induced deafness	6,820	16.8	6,110	15.9
Skin diseases	3,112	8.6	3,724	12.9
Asbestosis	2,284	11.5	1,915	9.6
Total (list system only)	77,896	345.2	67,570	336.3

is lower than the 1997 amount, but slightly higher than the 2002 amount (which was € 332.7 million).

3.4.2 Torts

32   In Section 3.1 we noted that in Belgium civil liability of employers is generally excluded. The importance of tort law as one of the 'layers of compensation' will therefore be small. On the other hand, as in any system based on *risque professionnel*, there will be legal procedures concerning questions such as the work-related nature of the disability. According to Rauws (2001), however, this problem might be a little exaggerated, as in Belgium the definitions of 'industrial accident' and 'occupational disease' have been widened over the years (more tolerant towards employees) both in legal practice and in legislation.[61]

4. Germany

4.1 Overview[62]

33   Like in Belgium, the German compensation system considers industrial accidents and occupational diseases as a *risque professionnel*. The *Unfallver-sicherungsgesetz* (Industrial Injuries Insurance Act, UV) applies, which was promulgated in 1884. This is a comprehensive system of public insurance, paid for by employers and managed by the so-called *Gewerbliche Berufs-genossenschaften* (BGs), which are organised along industry sector lines, each one with its own autonomous administration. The German compensation sys-

---

[60] Sources: Prevent (*supra* fn. 43), 39 and FBZ, *Jaarverslag* (2003).

[61] Rauws (*supra* fn. 36), 115–116.

[62] For a more elaborate overview I again refer to the contributions by Engelhard and Hoop. See also H.J.W. van Dongen, Schets van de Regeling tot Schadevergoeding van Beroepsziekten in Nederland, België, Duitsland en het Verenigd Koninkrijk, in: M.G Faure/T. Hartlief (eds.), *Verzekering en de Groeiende Aansprakelijkheidslast* (1995), 103–108.

tem applies to all employees and their financial dependants, as well as to those with 'equal legal status'. It does not apply to public officials.

The level of compensation is related to the degree of incapacity for work and is limited in size; with respect to income losses a maximum is fixed at two-thirds of lost wages. In addition to income losses medical costs (including rehabilitation costs) are covered by the system. There are, however, almost no possibilities to take civil action against employer or colleagues, with the exception of cases involving intent and gross negligence. General risks of illness and personal injury, i.e. not being work-related, are compensated via the *Kranken-versicherung* (Health Insurance Act) and the *Rentenversicherung* (Pension Insurance Act).

Since the introduction of the UV in 1884, several changes (additions) have been made to the system, such as the inclusion of commuting accidents and occupational diseases in 1925, the inclusion of accidents at school in 1971 and the addition over the years of several occupational diseases to the list of officially recognised diseases. As regards changes in financing, in 1976 the employers' contributions were made dependent on the number of accidents in the individual company, in addition to the industry risk factor. Since 1997 the UV has been incorporated in the *Sozialgesetzbuch Buch VII* (Social Code, SGB).[63]

As mentioned above, central in the German system are the BGs: 34

> 'in case of occupational accidents, commuting accidents and occupational diseases, the German Berufsgenossenschaften as the statutory accident insurance institutions are responsible for the provision of all rehabilitation services. They control and co-ordinate medical treatment (medical rehabilitation) as well as reintegration into professional life (professional rehabilitation) and the social environment (social rehabilitation). In order to guarantee a basic standard of living during the period of rehabilitation, injury benefits or temporary allowances are granted by the BGs'.[64]

If the rehabilitation measures do not lead to the re-establishment of unrestricted participation in working life, a 'pension' will be paid to the insured persons. The amount of such a pension depends on the reduction in earning capacity, as will be explained in Section 4.4 below. The reduction in earning capacity must be at least 20 percent.

The duties of the BGs are regulated by law. These include the provision of rehabilitation services and managing the compensation system (as just discussed) and looking after prevention of occupational accidents and occupational diseases. The *Hauptverband der Gewerblichen Berufsgenossenschaften* (Central Federation of BGs, HVBG) represents the common interests of all its members. The HVBG also decides on the contribution rates to be paid and preventive measures to be taken by companies.

---

[63] Before 1997 the UV was included in the *Reichsversicherungsordnung* (State Insurance Code, 1911).
[64] HVBG website: http://www.hvbg.de, 9 May 2006.

4.2 Industrial Accidents

35 According to the European Health and Safety Database, the HVBG has collected data on both occupational accidents and diseases on a regular basis from 1978 to the present. Accessibility is restricted:[65]

> 'The yearly published statistics on occupational accidents and diseases are legally prescribed; they are based on the recording and reporting procedures for accident insurance. The analysis and publication of statistics on occupational accidents and diseases do not merely have financial goals. The overviews and special studies are designed for use in prevention, rehabilitation, education and advisory tasks of the associations. They only have a secondary function for epidemiological or social research. [...] The data are only available for Federation experts. In special cases data are made available for research outside the institution'.

36 The HVBG website contains some of this statistical information on accidents at work. Table 4.1 shows the number of reportable accidents at work (*meldepflichtige Arbeitsunfälle*)[66] and the number of fatal accidents at work in the period 1994–2004, as well as the number of new pensions (resulting from a reduction in earning capacity or fatal accident) in each year. Pre-1994 data are also available and even go back as far as 1950. However, one cannot directly compare the recent data with figures from previous years because of a change in the 'statistical basis used in accident insurance' in 1986 and the 'inclusion of the new federal states in eastern Germany' in 1991.[67] The figures in Table 4.1 indicate a decrease, to which we will come back to below.

37 Table 4.2 contains similar data on the number of commuting accidents (*Wegeunfälle*). As in Belgium, commuting accidents are registered separately from accidents in the workplace. Apparently the number of commuting accidents fell somewhat in the period 1994–2004 (although the HVBG itself speaks of a stabilisation). Figures presented in the HVBG annual report 2003 indicate a decrease also in the years before that (1965–1989). The report does not give any explanations for this.[68]

38 Table 4.3 shows the number of industrial accidents per 1,000 full workers and per 1 million hours at work. Again, the figures show a decrease in the incidence of industrial accidents in Germany. Pre-1994 data, available from the HVBG website as from 1965, show a similar trend. In 1965 the accident rate

---

[65] Source: European Health and Safety Database (HASTE): http://www.ttl.fi/internet/partner/haste, 9 May 2006. The HASTE website is published and kept up-to-date by the Finnish Institute of Occupational Health. See also C. Jacinto/J. Aspinwall, A Survey on Occupational Accidents' Reporting and Registration Systems in the European Union, [2004] *Safety Science* 42, 935.

[66] Employers are obliged to report to the competent BG all accidents entailing an incapacity to work for more than three days.

[67] HVBG, *BG Statistics: Figures and Long-Term Trends 2003* (2004), 5.

[68] HVBG, *Geschäfts- und Rechnungsergebnisse 2003*, 18–19

Table 4.1 Industrial Accidents (1994–2004)[69]

Year	Reportable accidents at work	Fatal accidents at work	Accidents at work: New pensions
1994	1,489,360	1,250	34,659
1995	1,415,381	1,196	34,464
1996	1,266,458	1,120	33,966
1997	1,221,530	1,004	28,135
1998	1,198,608	948	25,549
1999	1,185,382	977	24,338
2000	1,144,262	825	22,678
2001	1,060,625	811	21,354
2002	937,540	773	20,603
2003	871,145	735	19,646
2004	841,447	645	18,138

Table 4.2 Commuting Accidents (1994–2004)[70]

Year	Reportable commuting accidents	Fatal commuting accidents	New commuting accident pensions
1994	191,387	829	9,495
1995	205,925	808	9,489
1996	196,517	748	10,141
1997	179,734	735	8,359
1998	184,310	695	7,677
1999	187,559	747	7,365
2000	177,347	772	6,929
2001	176,420	669	6,510
2002	168,353	581	6,640
2003	158,301	604	6,608
2004	151,330	497	6,272

per 1,000 full workers was 118.62, which is four times as high as the accident rate in 2004.[71] The decline of heavy industry and the closing of coal mines in Germany might be some of the explanatory factors for this. To what extent the decrease in the accident rate – and the 'new pensions' rate – can be credited to successful prevention efforts by the BGs is unclear.

---

[69] Source: Hauptverband der gewerblichen Berufsgenossenschaften (HVBG).
[70] Source: Hauptverband der gewerblichen Berufsgenossenschaften (HVBG).
[71] HVBG (*supra* fn. 69), 16–17. Unfortunately reasons for this strong decrease are not given.

Table 4.3 Reportable Accidents at Work: Accident Rates (1994–2004)[72]

Year	Accidents at work per 1,000 full workers	Accidents at work per 1 million hours at work
1994	50.13	31.93
1995	46.68	29.73
1996	40.49	26.64
1997	39.57	25.86
1998	39.38	25.41
1999	38.72	24.82
2000	37.10	24.09
2001	34.51	22.56
2002	32.45	21.21
2003	29.37	19.19
2004	27.85	17.63

4.3 Occupational Diseases

39   Data on the number of claims for recognition (*Anzeigen auf Verdacht einer Berufskrankheit*) and recognised cases (*Anerkannte Berufskrankheiten*) are again provided by the HVBG and are included in table 4.4 below. The last column gives information on the number of new pensions. Again, data are provided for the period 1994–2004. More (earlier) data are available but cannot be directly compared to the figures presented in Table 4.4 for reasons mentioned in the previous section.

Table 4.4 Occupational Diseases: Claims for Recognition (1990–2004)[73]

Year	Suspected cases (claims for recognition)	Recognised cases: Total	Recognised cases: Of which new pensions
1994	83,847	19,419	6,432
1995	78,429	21,886	6,705
1996	82,349	21,985	7,076
1997	77,310	21,187	6,983
1998	74,470	18,614	5,691
1999	72,722	17,046	5,309
2000	71,172	16,414	4,901
2001	66,784	16,888	5,189
2002	62,472	16,669	5,138
2003	56,900	15,758	4,799
2004	55,869	15,832	4,748

[72] Source: Hauptverband der gewerblichen Berufsgenossenschaften (HVBG).
[73] Sources: Eurogip (*supra* fn. 45), 30, HVBG (*supra* fn. 67), 55 and HVBG website (*supra* fn. 64, consulted 8 November 2005). The data provided in this table concern wage-earners of industry, commerce, services and some of the self-employed persons.

Table 4.5 presents the number of occupational diseases *per 100,000 insured*   40
persons as well as the ratio of recognised versus notified diseases.

Table 4.5 Occupational Diseases: Numbers per 100,000 Insured Persons (1990–2000)[74]

Year	Number of claims per 100,000 insured persons	Recognised ODs/ notified ODs ratio	Number of recognised ODsper 100,000 insured persons
1990	192	18.3%	35
1991	181	17.1%	31
1992	219	16.6%	36
1993	281	19.4%	54
1994	256	23.2%	59
1995	235	27.9%	66
1996	249	26.7%	66
1997	230	27.4%	63
1998	224	25.0%	56
1999	216	23.4%	51
2000	211	23.1%	49

These figures show that since 1993 there has been a downward trend in the
number of claims for recognition per 100,000 insured persons. According to
the European Forum of Insurances against Accidents at Work and Occupational Diseases, this '*is mainly due to the success of prevention and to changes in
the economic structure, e.g. the reduction in the number of coal mines*'.[75] The
low value of the recognition/notification ratio can be related to the fact that
there are no restrictions in the German reporting procedure. Also, there are no
time limits for making a report. This leads to many reports that have no chance
of being accepted.

As far as the diseases leading to the greatest number of claims for recognition   41
is concerned, the European Forum presents data for the year 2000. In that year
most claims in Germany were related to the category 'Skin diseases (except
cancer)', followed by the categories 'Hearing loss', 'Back diseases', 'Asbestosis and pleural plaques' and (ranking fifth) 'Allergic respiratory diseases'.[76]

4.4 Amounts of Compensation Received

In Germany compensation for income losses is paid out by one source only: 42
the BGs. This is very different from the situation in the Netherlands (as described in Section 2.4.1) where various compensation sources exist side by
side. In the case of Germany, one could therefore begin by listing the total expenses by the BGs. In 2003 a total of € 7.6 billion was spent, of which € 2.6

---

[74] Source: Eurogip (*supra* fn. 45), 30.
[75] Eurogip (*supra* fn. 45), 15.
[76] Eurogip (*supra* fn. 45), 17.

billion on rehabilitation, € 5.0 billion on pensions and other cash benefits and
€ 0.7 billion on prevention. Numbers for the years before that are also available and they show that there has been a slight increase in expenditure on
compensation over the years. However, this increase is probably caused mainly by inflation, because the 'expenditure per € 100 of wages' has remained
constant (between 1.11 and 1.13).

The calculation of the pensions payable to insured persons and surviving
dependents is based on the gross annual earnings (*Jahresarbeitsverdienst*) and
the so-called MdE, which stands for reduction of earning capacity (*Minderung
der Erwerbsfähigkeit*). *E.g.*, if MdE is 100 percent, the insured person will receive a 'full pension', which amounts to two thirds of his/her (former) gross
annual earnings.

43   As stated in Section 4.1, tort law does not play an important role in the German
compensation system, with the possible exception of cases involving intent
and gross negligence from the side of the employer. The HVBG does not have
information on these cases.

5.   Great Britain[77]

5.1   Overview[78]

44   In Great Britain both tort law and social security are important when it comes
to seeking compensation for income losses caused by industrial accidents and
occupational diseases. Hence one might argue that the British compensation
system resembles the Dutch one more than it does the Belgian or German systems. According to the Employers' Liability Act of 1969, employers[79] are
obliged to take out liability insurance against the costs of compensation for
employees who are injured or made ill at work through the fault of the employer. These costs of compensation may include loss of earnings, health care
costs and 'pain and suffering'. Reasons for the obligation to insure are presented in a recent UK Government report:[80]

> 'It provides greater security: to firms against costs which could otherwise
> result in financial difficulty; and to employees that resources will be
> available for compensation even when firms have become insolvent. It
> supports the right of employees injured through their employer's negli-
> gence to be fairly compensated – 'access to justice' – and the responsibil-
> ity of employers to fund the costs of their negligence – 'polluter pays'.

In addition, the Government believes that making employers fund the cost of
their negligence might have benefits as far as the prevention of accidents is

---

[77] Following most data sources, the statistics presented in this section refer to England, Scotland
and Wales.
[78] For a more elaborate overview I again refer to the contributions by Engelhard and Hoop.
[79] Except for nationalised industries, the police and local authorities.
[80] Department for Work and Pensions, *Review of Employers' Liability Compulsory Insurance:
First Stage Report* (2003), 14.

concerned. Finally, the interference of insurers might lead to improvements in risk management, if provisions about this are included in the insurance contract.

Prior to 1972, the year the Employers' Liability Act became effective, claims   45
against employers often failed if the employer was not insured, because insufficient funds existed. Now failure to insure is regarded as a criminal offence. Currently the Government is reviewing the Employers' Liability Act, as apparently premiums that have to be paid by employers to liability insurers have risen significantly in recent years.[81] One of the suggestions put forward by the Government is to make premiums more risk-related.[82] Namely, frequent complaints by smaller businesses suggest that premiums do not reflect their good health and safety practices and claim records, but are instead set by insurers on the basis of a standard book rate.

As far as social security is concerned, the National Health Service (NHS) is   46
well-known. The NHS covers general health care losses and is financed mainly by public funds. However, the *National Insurance* covers the financial risk of injury arising in and out of the course of employment, financed by contributions of employers and employees.[83] Benefits consist of flat-rated payments for income loss and medical treatment. In cases of disablement caused by an industrial accident or occupational disease, the Industrial Injury Disablement Benefit (IIDB) applies. Contrary to the employers' liability system, IIDB does not involve fault being established.[84] Moreover, compensation of income losses via IIDB does not depend on the actual income loss but on the degree of incapacity, and it is not payable for the first 90 days after an accident. This of course means that IIDB does not provide for full loss of earnings. With respect to sick leave, employees will be paid statutory sick pay for periods between 4 days and 28 weeks, which is financed by employers.

IIDB and employers' liability are separate systems. However, the 2003 Gov-   47
ernment report mentioned above explains that there is interaction between the two:[85]

> 'IIDB provides a safety net for employees suffering accident, injury or disease at work who choose not to pursue a claim against their employer

---

[81] See Department for Work and Pensions (*supra* fn. 80) and Department for Work and Pensions, *Employers' Liability Compulsory Insurance: Second Stage Report* (2004). The recent premium increases have been driven by a cyclical change in the insurance market, increases in legal costs and (possibly) uncertainty over long-tail risks.

[82] For more information see the original reports.

[83] More generally: The National Insurance offers benefits such as the Incapacity Benefit, Retirement Pension, Jobseeker's Allowance, Maternity Allowance and Bereavement Benefit. Some of these benefits are dependent on the level of the so-called 'voluntary contributions' made to the system. See, *e.g.*, http://www.adviceguide.org.uk (consulted 9 November 2005).

[84] In Great Britain the employer's civil liability, which is based on fault, must be shown by the claimant.

[85] Department for Work and Pensions (*supra* fn. 80), 18.

and for cases where liability cannot be established. Pursuing a case for negligence against an employer can be expensive and time consuming. Proving negligence can be difficult especially where the injury or disease took place some time ago. In some cases, the employer may no longer be in business. In some instances IIDB is also the means of support for employees whilst they pursue the more lengthy employers' liability claims process'.

5.2 Industrial Accidents

48   The Health and Safety Executive (HSE) has collected data on industrial accidents in Great Britain since 1986. Together with the Health and Safety Commission (HSC), which is sponsored by the Department for Work and Pensions, the HSE is responsible for the regulation of most risks to health and safety arising from work activity. According to the European Health and Safety Database:[86]

> 'Data on occupational accidents is compiled by the Health and Safety Executive which produces statistics on occupational accidents in England, Wales and Scotland reported by employers and the self-employed to the enforcing authorities under the Reporting of Injuries, Diseases and Dangerous Occurrences Regulations 1985 (RIDDOR 85). The annual statistics principally produce data on three categories of accidents: fatal accidents, those causing defined major injuries[87] and those resulting in an absence from work for more than three days. [...] There is no access to the databases for people outside the HSE'.

On its website and in various reports the HSE presents key figures on work injuries in Great Britain.[88] As far as recent figures are concerned: There were 168 fatal injuries, 30,689 major injuries and 131,017 over-3-days injuries in 2003/2004 reported by employers. About 2.0 million people were suffering from 'an illness which they believed was caused or made worse by their current or past work'. Overall, the HSE states that an estimated 35 million working days were lost (1.5 per worker), of which 28 million due to work-related ill health and 7 million due to workplace injury.

49   Table 5.1 presents historical injury figures (i.e. injuries resulting from industrial accidents) for the private sector. The figures indicate a downward trend in the number of reported *fatal injuries*, although this trend is definitely not linear. With respect to *major injuries*, we note a big change around 1996 that has been caused by the replacement of RIDDOR 85 by new reporting regulations (RIDDOR 95) in 1995. In recent years the number of reported major injuries

---

[86] *Supra* fn. 65 (consulted 9 November 2005).
[87] Examples of major injuries include amputations, dislocations (of shoulder, hip, knee, spine), fractures (except to fingers, thumbs or toes) and other injuries leading to resuscitation or 24 hour admittance to hospital. See HSE, *Technical Note – Safety Statistics*, at http://www.hse.gov.uk (consulted 26 May 2005).
[88] See in particular Health and Safety Commission, *Health and Safety Statistics 2004/05* (2005).

has increased almost every year, according to the HSE mainly in service industries. The number of *over-3-days injuries* has remained fairly constant as far as employees are concerned (except for the preliminary 2004/05 number), while concerning the self-employed the drop in 1996/97 probably resulted again from the changes in reporting regulations. Public sector data are available too (with the exception of over-3-days injuries) but these are not provided here.[89] Also, many HSE reports include figures presenting the number of injuries by industry and by kinds of accident.

Table 5.1 Reported Occupational Injuries in the Private Sector 1986/87 – 2004/05[90]

Year[91]	Employees			Self-employed		
	Fatal	Major	Over 3 days	Fatal	Major	Over 3 days
1986/87	355	20,695	159,011	52	690	1,029
1987/88	361	20,057	159,852	84	867	1,169
1988/89	529	19,944	163,119	80	1,152	1,503
1989/90	370	20,396	165,244	105	1,310	1,865
1990/91	346	19,896	160,811	87	1,326	2,077
1991/92	297	17,597	152,506	71	1,101	1,832
1992/93	276	16,938	141,147	63	1,115	2,136
1993/94	245	16,705	134,928	51	1,274	2,531
1994/95	191	17,041	139,349	81	1,313	2,869
1995/96	209	16,568	130,582	49	1,166	2,394
1996/97	207	27,964	127,286	80	1,356	2,282
1997/98	212	29,187	134,789	62	815	984
1998/99	188	28,368	132,295	65	685	849
1999/00	162	28,652	135,381	58	663	732
2000/01	213	27,524	134,105	79	630	715
2001/02	206	28,011	129,655	45	929	917
2002/03	183	28,113	128,184	44	1,079	951
2003/04	168	30,689	131,017	68	1,283	1,114
2004/05p	169	30,213	120,346	51	1,246	1,135

[89] Surprisingly though, the number of fatal injuries in the public sector has since 1996/97 been larger than the total number of fatal injuries in the private sector. In 2003/04 there were 374 fatalities in the public sector. Before 1996/97, i.e. under RIDDOR 85, the number of fatalities in the public sector was much lower. The number of 'non-fatal injuries' leading to hospitalisation in the public sector in 2003/04 was 13,679. For more information: see HSE website.

[90] Sources: HSE, *Historical Injury Figures* (2005) and Health and Safety Commission (*supra* fn. 88). Numbers for 2004/05 are preliminary.

[91] Figures are based on a planning year 1 April–31 March. 1987/87–1995/96 reported under RIDDOR 85; thereafter reported under RIDDOR 95. The 2004/05 data are provisional. Some figures for 1981–1985 are also available, but on a calendar year basis and reported under the Notification of Accidents and Dangerous Occurrences Regulations (NADOR) 1980.

50   Again, to better allow for year on year comparison, historical incidence rates
      are provided. Table 5.2 shows rates of injuries per 100,000 members of the
      workforce. There was a general downward trend in the rate of *fatal injuries* of
      workers in the 1990s, but it has risen twice since then. The fact that the fatal
      injury rate is now higher for the self-employed than for employees, according
      to the HSE reflects the fact that proportionally more self-employed people
      than employees work in the higher risk industries of agriculture and construc-
      tion. And of course the rate of fatal injuries of the self-employed is more sus-
      ceptible to change because there are much less self-employed people than em-
      ployees.[92] The recent rise in the rate of *major injuries* of employees is, as
      mentioned before, mainly caused in service industries. Before that, from 1997/
      98 to 2000/01 this rate fell steadily, also for the self-employed. The rate of *over-
      3-days injuries* gradually decreased from the mid 1990s, but rose again once in
      2003/04 (and has risen continuously since 2001/02 for the self-employed).

Table 5.2 Reported Occupational Accidents in the Private Sector: Incidence Rates (per
100,000)[93]

Year[94]	Employees			Self-employed		
	Fatal	Major	Over 3 days	Fatal	Major	Over 3 days
1986/87	1.7	99.1	761.1	2.0	26.9	40.1
1987/88	1.7	94.0	748.9	3.0	31.0	41.4
1988/89	2.4	91.4	747.7	2.7	39.4	51.4
1989/90	1.7	91.8	743.4	3.3	41.2	58.6
1990/91	1.6	89.9	726.5	2.7	41.2	64.5
1991/92	1.4	81.7	708.5	2.3	35.9	64.5
1992/93	1.3	80.3	669.0	2.0	35.8	68.5
1993/94	1.2	79.3	640.2	1.6	40.6	80.7
1994/95	0.9	80.4	657.2	2.5	40.4	88.4
1995/96	1.0	77.1	607.4	1.5	36.0	73.8
1996/97	0.9	127.5	580.1	2.3	38.4	64.6
1997/98	0.9	127.6	589.2	1.8	23.3	28.1
1998/99	0.8	121.7	567.3	1.9	20.3	25.2
1999/00	0.7	116.6	550.9	1.7	19.7	21.8
2000/01	0.9	110.2	536.9	2.4	19.2	21.8
2001/02	0.8	110.9	513.5	1.3	27.8	27.5
2002/03	0.7	111.1	506.5	1.3	32.3	28.4
2003/04	0.7	120.4	514.2	1.8	33.9	29.5
2004/05p	0.7	117.7	469.0	1.3	32.9	29.9

[92] HSE, *Health and Safety Statistics Highlights 2003/04* (2004), 4.
[93] Sources: HSE, *Historical Injury Figures* (2005) and Health and Safety Commission (*supra*
fn. 88). Numbers for 2004/05 are preliminary.
[94] *Supra* fn. 91.

5.3   Occupational Diseases

The Reporting of Injuries, Diseases and Dangerous Occurrences Regulations   51
(RIDDOR) require employers to report all cases of diseases listed in a *Schedule to the Regulations*, if they receive a doctor's written diagnosis and the affected employee's current job involves the work activity specifically associated with the disease. The list of diseases (and their associated occupational
conditions) was revised and extended in 1995. According to the HSE, the
wide publicity given to RIDDOR 95 at the time is likely to have encouraged
reporting of diseases. However, the HSE also gathers statistics on occupational diseases itself through its Surveys of self-reported ill health (SWI), mainly
because there is substantial under-reporting under RIDDOR.

The HSE reported 2.2 million cases of 'work-related ill health' in Great
Britain in 2003/04[95], which equates to 5.2% of people who have ever worked
in Great Britain. These numbers are based on SWI03/04. Four similar household surveys were carried out by the HSE in 1990, 1995, 1998/99 and 2001/
02, respectively. Comparison between these surveys suggests that:[96]

> 'the overall rate of self-reported work-related illness prevalence has
> fallen since 1990. In 1990 and 1995 the rates were similar. More recently
> they have fluctuated, but are still lower than in 1990 and 1995'.

However, it should be noted that the numbers are estimates. Moreover, they
depend on lay people's perceptions of medical matters.

The most commonly reported disorders in 2003/04 were musculoskeletal dis-   52
orders (bone, joint or muscle problems) followed by the categories 'Stress, depression or anxiety', 'Breathing or lung problems' and 'Hearing problems'. In
its reports on work-related illnesses, the HSE focuses primarily on individual
categories of diseases and on sector/industry analyses, as this is much more
interesting from the point of view of prevention. It is therefore much more difficult to find information on the general prevalence of occupational diseases in
Great Britain.

5.4   Amounts of Compensation Received

In Section 5.1 we explained that income losses resulting from industrial acci-   53
dents and occupational diseases in the UK are compensated via the Employers' Liability Act – if a tort claim is successful – and via the Industrial Injury
Disablement Benefit (IIDB) in social security. Of course, compensation via
social security is limited and standardised, but it does not involve fault being
established. A claim can generally be met if it can be established that the injury or illness is work related and the degree of disability (judged by a medical
examination) is at least 14%. An IIDB benefit provides for the following: a
disablement pension, which is linked to the degree of disability; a constant at-

---

[95]   As stated in the previous section, the estimation for 2004/05 is lower: 2.0 million.
[96]   *Supra* fn. 92, 6. For more detailed statistics and in-depth analysis, see HSE, *Self-Reported
       Work-Related Illness in 2003/04: Results from the Labour Force Survey* (2005).

tendance allowance (CAA) to distinguish between full- and part-time workers, which is paid at four different rates; an unemployability supplement; and an allowance that is only paid out in severe cases where disablement is 100%. The Department for Work and Pensions (2003) gives some examples:[97]

> 'Disability allowances are set at fixed weekly amounts such as £79.03 for 70% disablement versus £33.87 for 30% disablement. Similarly, CAA is set at fixed amounts such as £90.40 for exceptional rate versus £22.60 for part-time rate. Unemployability supplement is £69.75 with additions for early incapacity'.

If victim-employees successfully sue their employer, damages include compensation for loss of earnings, health care costs and pain and suffering. Contributory negligence on the part of the employee would reduce the damages.[98]

54   As far as the *number* of claims is concerned, there is little information. Engelhard[99] writes the following:

> 'Reputedly, nowadays [NP: 1999] the English face, at the outside calculation, annually 115,000 tort claims for industrial injuries of which about 78% seemingly result in some payment of compensation. At the very least 3,000 of those claims are said to be for occupational diseases (RSI, stress and alike) of which 57% seem to result in compensation payments. [See Cane, *Atiyah's accidents, compensation and the law* (1999), 179, who] makes mention of the figures but argues 115,000 is probably now rather too high given the fall in the last thirty years in the numbers of people killed and injured at work'.

55   In 2004 the HSE's Economic Advisers Unit published an interim report on the 'Costs to Britain of Workplace Accidents and Work-Related Ill Health', which includes estimations of employers' liability costs. The report provides some broad indications of the costs to individuals, employers and society of work-related incidents in Great Britain, based on 2001/02 data. Cost categories to individuals include lost earnings, extra expenditure when absent and 'pain, grief and suffering'. Costs to employers include compensation (via employers' liability), sick pay (excluding statutory sick pay) and others such as administration, insurance and recruitment costs. Costs to society consist of costs of medical treatment (NHS), loss of output, HSE investigation costs, 'pain, grief and suffering' and others.[100] The estimations presented in the report are as follows[101]

---

[97] Department for Work and Pensions (*supra* fn. 80), 16.
[98] Department for Work and Pensions (*supra* fn. 80), 17.
[99] See the contribution by Engelhard to this book.
[100] A figure representing the different cost categories can be found in the Economic Advisers Unit report, which is available from the HSE website (*supra* fn. 87, consulted 9 May 2006).
[101] Source: HSE. The original tables also include non-injury accident costs, representing the cost of damage to materials, machinery and property.

Table 5.3 Costs of Workplace Incidents (× bln. £)

	Injury	Ill Health
Costs to individuals	3.3 to 6.3	5.9 to 9.4
Costs to employers	1.0 to 1.1	1.5
Costs to society	5.9 to 10.7	11.3 to 17.3
Costs to the economy	3.2 to 6.2	7.6 to 11.6

The difference between 'costs to society' and 'costs to the economy' is that the latter excludes the costs of pain, grief and suffering. Incidentally, the HSE notes that these costs are especially very hard to measure. The very wide range of estimates in Table 5.3 once again shows that finding concrete data on amounts paid out under tort law and social security is, unfortunately, very difficult.

6. Concluding Remarks

In this paper we presented some first empirical data on the incidence and scope of work-related incidents in The Netherlands, Belgium, Germany and Great Britain. We found that one has to be careful in interpreting these data, as different definitions of industrial accidents and occupational diseases are used in the various countries. Moreover, in some cases there have been changes over time in reporting regulations (Great Britain) or in the presentation of data (Germany), while in others statistics have become available only recently (Netherlands). Nevertheless, we were able to at least present some figures on fatal accidents and serious accidents for all countries under review here, including historical data (with the exception of the Netherlands). We focused on private sector data. Most interesting are of course the accident rates. The figures presented in this paper indicate that in Belgium, Germany and Great Britain[102] the reported accident rates have shown a downward trend in recent years, both for fatal accidents and for serious accidents. Unfortunately, clear explanations for these decreases could not be given; *e.g.* we do not know yet if they result (also) from successful prevention efforts by the government. To some extent the decreases might have been caused by changes in the proportion of the labour force employed in high risk industries, or changes in other factors such as the proportion of flexible workers. There is one exception to the downward trend, namely the rate of serious accidents in Great Britain (major and over-3-days injuries), which has risen recently due to more reported accidents in the service industry. We also presented some data on the prevalence of occupational diseases in the four countries. Again it appears that the reported overall rate of occupational disease (notifications and recognitions) has fallen slightly in Belgium, Germany and Great Britain, although the picture may change considerably when individual categories of diseases are considered. In Germany, where the decrease is most apparent, the downward trend has been ascribed to the success of prevention and changes in the economic

56

---

[102] There are insufficient data regarding the situation in the Netherlands.

structure. We do not yet know if the downward trend in Belgium and Great Britain can be related to (minor) shifts in governance, prevention efforts or something else. However, we do know – and have pointed out repeatedly – that in all four countries under review there is a problem of under-reporting.

57   In addition to the above, we identified the sources of compensation for personal injury and illness in the countries concerned. We concentrated our discussion on the compensation for lost income but occasionally commented on compensation for health care costs and pain and suffering as well. Theoretically there are several possible 'compensation layers', such as social security, no-fault compensation funds, private insurance and tort law, and different combinations are used in each of the countries. In the Netherlands tort law is relatively important, next to social security and private insurance. We presented some key figures on Dutch social security (historical WAO data with respect to disability benefits and historical *Ziektewet* data with respect to sickness benefits) and on the estimated (general) number of claims. However, for an empirical (regression) analysis of the effects of shifts in governance in the Netherlands[103] more and better data are definitely needed, although unfortunately very hard to find. In Belgium the role of tort law is much smaller. Instead there is a no-fault insurance system paid for by employers collectively, which is part of the social security system. Employers are obliged to take out insurance in order to cover the risk of industrial accidents. Compensation of lost income is standardised in this system. We presented some data on the amounts of compensation paid out by liability insurers, and it appears these amounts have been rising recently because of the increased costs of serious accidents. Again, more information is needed to make firm statements. Compensation for occupational diseases is also standardised in Belgium – it depends on the employee's disability percentage – and is received directly from the so-called Occupational Diseases Fund. Some data on the total amount of damage payments paid by this fund can be found in Section 3.4. In Germany compensation for work-related income losses and medical costs are paid out by the *Berufsgenossenschaften* (BGs), who manage the comprehensive public insurance system funded by employers. As in Belgium, amounts are limited in size (they are related to the degree of incapacity for work) and tort law plays only a minor role. General risks of illness and personal injury are compensated via medical insurance and pension insurance. We presented some recent figures on the expenses by the BGs, but more historical information on the pensions paid out to victims of workplace incidents is still needed. In the UK the potential compensation layers are tort law (combined with insurance) and social security. Employers are mandatorily insured against the costs of compensation (income loss, healthcare costs and pain and suffering) for employees who are victims of workplace incidents. With respect to social security we discussed the IIDB (Industrial Injury Disablement Benefit), which is standardised. However, (again) more historical data are needed for an analysis of possible shifts in governance in the UK.

---

[103]   *E.g.*, some minor shifts in social security occurred in the early 1990s and 2000s. See Section 2.1.

We mentioned a number of times that it is difficult to interpret statistics on   58
workplace accidents, because various definitions are used in various countries.
For an analysis of shifts in governance within one single country, this does not
necessarily have to be a problem. However, if one would like to compare several countries[104] – *e.g.* in order to analyse the effectiveness of different compensation systems – this may create severe problems. Eurostat, the Statistical
Office of the European Communities, is working on ESAW (European Statistics on Accidents at Work), a programme to give consistency to industrial accident statistics in the EU. The idea is to harmonise definitions of workplace
accidents in co-operation with EU member states. Some notable differences
between countries still remain as far as accident statistics are concerned: *e.g.*
not all countries include road traffic accidents and commuting accidents in
their statistics, and some countries have only limited information on public
sector accidents and on the self-employed. We also noted that in some countries the definition of 'serious' accidents is based on at least one working day
lost (Netherlands, Belgium) while in other countries this threshold is three
working days (Germany, Great Britain). The data collected by Eurostat cover
only the nine branches of industry that are common to all member states.[105]
Altogether, the ESAW data differ quite largely from the figures presented in
this paper.[106] Moreover the problem remains that some countries, such as the
Netherlands, do not have much historical data. For all these reasons we did not
present the ESAW data here, however interesting they may be. Another Eurostat project deals with occupational diseases (European Occupational Diseases
Statistics, EODS), on the basis of a methodology developed from 1997.[107]

To conclude, there remains a lot to be done in order to (empirically) analyse   59
the compensation and prevention effects of shifts in governance in the four
countries under review here. With respect to compensation, concrete figures
regarding amounts paid out by insurers and social security agencies to victims
of workplace accidents are lacking. Information on amounts paid out via tort
law could not be found at all, with the exception of some (too) general data
from Great Britain. The empirical data presented in this paper are indeed first
empirical data; they are certainly helpful in directing any further research, but
more and better (more specific) data are needed in order to make firm conclu-

---

[104] We stated in the introduction to this paper that this is not the central focus of this paper, but
may be the subject of further research.

[105] HSE, *Statistics of Workplace Fatalities and Injuries in Great Britain: International Compari-
sons 2000* (year of publication unknown). The nine branches are agriculture, manufacturing,
utilities, construction, retail and wholesale distribution, hotels and restaurants, transport,
financial services and real estate business activities.

[106] Therefore it is proposed that member states provide Eurostat with accident data which:
exclude commuting accidents; are based on the 3 day criterion for absence; identify road traffic accidents; cover the same industries; and cover all types of employment. See HSE (*supra*
fn. 105), 9.

[107] For statistics and more information on the Eurostat projects see http://europa.eu.int/comm/
eurostat and especially Eurostat, *European Statistics on Accidents at Work (ESAW): Methodology – 2001 Edition* (2001) and Eurostat, *European Occupational Diseases Statistics (EODS):
Phase 1 Methodology* (2000).

sions. Finally, we have to keep in mind that even the relatively simple statistics on the incidence of accidents and diseases should be handled with care, because of the different definitions used in various countries and because of changes over time in reporting regulations and the presentation of data (Great Britain, Germany). This will of course also complicate a further analysis of the deterrent effects of particular compensation systems. The best – and most interesting – way to circumvent this problem would be to concentrate only on particular sectors or, as far as occupational diseases are concerned, on specific types of work-related illness (*e.g.* hearing loss or asbestosis).

Bibliographic References

CANE, P., *Atiyah's Accidents, Compensation and the Law*, London/Edinburgh/Dublin: Butterworths, 1999.

DEWEES, D., DUFF, D. & TREBILCOCK, M., *Exploring the Domain of Accident Law: Taking the Facts Seriously*, Oxford University Press, 1996.

FAURE, M.G., HARTLIEF, T. & PHILIPSEN, N.J., Funding of Personal Injury Litigation and Claims Culture: Evidence from the Netherlands, *Utrecht Law Review*, Vol. 2, No. 2, 2006, pp. 1–21.

FAURE, M.G. & VAN DEN BERGH, R., Restrictions of Competition on Insurance Markets and the Applicability of EC Antitrust Law, *KYKLOS*, Vol. 48, 1995, pp. 65–68.

JACINTO, C. & ASPINWALL, E., A Survey on Occupational Accidents' Reporting and Registration Systems in the European Union, *Safety Science*, Vol. 42, 2004, pp. 933–960.

KLOSSE, S., Schadevergoeding via Sociale Zekerheid en Aansprakelijkheidsrecht: Communicerende Vaten?, in: FAURE, M. & HARTLIEF, T. (eds.), *Schade door Arbeidsongevallen en Beroepsziekten*, Den Haag: Boom Juridische Uitgevers, 2001, pp. 1–18.

POPMA, J.R. & VENEMA, A., Occupational Accidents in the Netherlands, in: SMULDERS, P. (ed.), *Worklife in the Netherlands*, Hoofddorp: TNO, 2006, pp. 165–179.

RAUWS, W., Financiering van Schade Veroorzaakt door Arbeidsongevallen en (Nieuwe) Beroepsziekten: België als Wenkend Voorbeeld?, in: FAURE, M. & HARTLIEF, T. (eds.), *Schade door Arbeidsongevallen en Beroepsziekten*, Den Haag: Boom Juridische Uitgevers, 2001, pp. 109–129.

SHAVELL, S., Liability for Harm versus Regulation of Safety, *Journal of Legal Studies*, Vol. 13, 1984, pp. 357–374.

VAN DE GOOR, A., *Effects of Regulation on Disability Duration*, Utrecht University, Thesis Publishers Amsterdam, 1997.

VAN DONGEN, H.J.W., Schets van de Regeling tot Schadevergoeding van Beroepsziekten in Nederland, België, Duitsland en het Verenigd Koninkrijk, in: FAURE, M.G. & HARTLIEF, T. (eds.), *Verzekering en de Groeiende Aansprakelijkheidslast*, Deventer: Kluwer, 1995, pp. 87–113.

VAN MIERLO, J.G.A., Economische Inschatting van de Evolutie in de Kosten, in: FAURE, M.G. & HARTLIEF, T. (eds.), *Verzekering en de Groeiende Aansprakelijkheidslast*, Deventer: Kluwer, 1995, pp. 221–282.

WETERINGS, W.C.T., *Vergoeding van Letselschade en Transactiekosten: Een Kwalitatieve en Kwantitatieve Analyse*, Deventer: W.E.J. Tjeenk Willink, 1999.

Prevention and Compensation of Work Injury in the United States: An Overview of Existing Empirical Evidence

*N.J. Philipsen**

1. Introduction

This paper has been composed within the framework of the 'Shifts in Governance' project. One of the main objectives of this project is to identify shifts from one personal injury compensation system to another (e.g. from public funding to civil law), as well as shifts within a particular compensation system. In that respect a number of countries and various domains of accident law (notably work injury, medical negligence and environmental damage) have been analysed. As far as work injury (industrial accidents and occupational diseases) is concerned, we have considered the Netherlands, Belgium, Germany and Great Britain. Other 'Shifts in Governance' papers in this domain have dealt with, respectively, the legal analysis of shifts in governance and a presentation of first empirical findings in these four countries.[1]   1

In this context it is very useful to briefly review past empirical research on personal injury compensation systems. Most of this research has been conducted in the United States, where many books and articles have been published especially on the compensation of victims of traffic accidents and medical negligence. I will first address, briefly, some of the more general arguments put forward in the American literature concerning the goals (objectives) of tort law and the extent to which these goals have been achieved according to various authors. This literature focuses on deterrence and compensation effects of tort   2

* Maastricht University, METRO Institute for Transnational Legal Research, PO Box 616, 6200 MD Maastricht, The Netherlands, E-Mail: niels.philipsen@facburfdr.unimaas.nl. This paper was written during a visiting scholarship at the University of Illinois at Urbana-Champaign in the early Summer of 2005. I would like to thank Prof. T.S. Ulen and Prof. T. Ginsburg for valuable suggestions and support and Prof. M.G. Faure for comments on a first draft of this paper.
[1] See the contributions by Engelhard, Hoop and Philipsen to this book.

law and will be the subject of Section 2. That section also contains a short discussion of alternative compensation systems, such as no-fault compensation funds, insurance and social security.

Then I will focus on empirical findings regarding prevention and compensation of industrial accidents and occupational diseases specifically. The existing literature on this topic mainly discusses the functioning of the workers' compensation system and the shift from tort law to workers' compensation in the early twentieth century. Some of the most interesting findings of this literature will be presented in Section 3. Section 4, finally, contains some concluding remarks.

2.   *Objectives of Tort Law*

3   When analysing shifts in compensation for industrial accidents in European countries, one can find valuable empirical information in the already existing American literature on the effects of various compensation systems, notably on the effects of the tort system. In this section I will briefly address the ongoing debate in this literature about the objectives of tort law (2.1). Distinction is generally made between deterrence and compensation perspectives on tort law.[2] Attention will be given also to alternative (or additional) personal injury compensation systems, such as no-fault compensation funds, insurance and social security (2.2).

2.1   Tort Law

2.1.1   Deterrence

4   Economists generally argue that a main objective of tort law is deterrence of wrongful and dangerous behaviour, while lawyers tend to stress its compensation aspect. The deterrence argument is straightforward: if there is a (credible) threat of a liability suit, potential injurers will behave more carefully, which will, *ceteris paribus*, result in a lower accident probability.[3] Social security and no-fault compensation funds do not have this deterrent effect, unless they are backed by (safety) regulation or unless financial contributions to the social

---

[2] Dewees, Duff and Trebilcock (1996) discuss three 'competing normative perspectives' on tort law, being deterrence, compensation and corrective justice. According to the latter perspective, the objective of tort law is to correct past injustices. I will not discuss the corrective justice perspective here. See, in particular D. Dewees/D. Duff/M. Trebilcock, *Exploring the Domain of Accident Law: Taking the Facts Seriously* (1996), 8–9.

[3] Here I implicitly assume that only the potential injurer can influence the accident risk (unilateral accident setting). In a bilateral accident setting, both the level of care taken by the injurer and the potential victim must be taken into account. For basic literature on the economic analysis of tort law, i.e. dealing with unilateral and bilateral accidents, activity level, strict liability versus negligence, etc., I refer to R. Cooter/T. Ulen, *Law & Economics* (4th edn., 2004), 307–349, and R.A. Posner, *Economic Analysis of Law* (6th edn., 2003), 167–213. See also the classic paper by S. Shavell, Strict Liability versus Negligence [1980] *Journal of Legal Studies* (JLS), 1 *et seq.*

security system or fund are made dependent on factors relating to the accident risk.[4]

Whether, in practice, tort law is indeed an efficient instrument to deter wrong-　5 doing is still highly debated in the literature. Some scholars suggest that the tort system may lead to a 'claims culture'[5] and to overdeterrence of risky activities, while others claim that tort law does not deter unduly dangerous conduct at all. Galanter (1998), for example, criticises what he calls the 'jaundiced view' of American civil justice: 'a set of beliefs and prescriptions about the legal system based on the perception that people are suing each other indiscriminately about the most frivolous matters, and juries are capriciously awarding immense sums to undeserving claimants'.[6] He presents numerous examples of atrocity stories, media distortion and rent-seeking behaviour by 'the elite' (politicians, big corporations, etc.) and argues that the jaundiced view of the American legal system has not been supported well empirically and is to a large extent based on contemporary legends. Galanter continues by referring to a large body of empirical literature on the civil justice system, which generally shows that the number of tort claims is much lower than is often claimed and that damage awards are not that excessive.[7] Much of this empirical research, which was later labelled the 'reassuring view' by Schwartz[8], has been conducted by (left-wing) law and society scholars, such as Galanter himself.

Schwartz (2002), in an interesting paper about empiricism and tort law, takes　6 an intermediate position. First he notes that both Saks (1992) and Galanter (1996) have already written long 'meta-review papers' on the operation of the U.S. tort litigation system.[9] Saks and Galanter are both supporters of the reassuring view and hence suggest in their respective studies that the American tort system exhibits a significant underenforcement of legally valid claims, instead of an excessive number of claims.[10] Schwartz, however, criticises them for misinterpreting a study conducted by the RAND Corporation's Institute

---

[4]　On the criteria that determine the choice between safety regulation and tort law, see S. Shavell, Liability for Harm versus Regulation of Safety, [1984] *Journal of Legal Studies* (JLS), 357 *et seq.*

[5]　Faure, Hartlief and Philipsen discuss the supposed 'claims culture' in the Netherlands. See M.G. Faure/T. Hartlief/N.J. Philipsen, Funding of Personal Injury Litigation and Claims Culture: Evidence from the Netherlands, [2006] *Utrecht Law Review*, Vol. 2, 1–21.

[6]　M. Galanter, An Oil Strike in Hell: Contemporary Legends about the Civil Justice System, [1998] *Arizona Law Review* (ArizLRev), 717.

[7]　M. Galanter, [1998] ArizLRev, note 14 on pp. 721–722.

[8]　G.T. Schwartz, Empiricism and Tort Law, [2002] *University of Illinois Law Review* (UIllLRev), 1074. Furthermore, Schwartz does not use the term 'jaundiced view' (as used by Galanter) but instead refers to the 'alarmist view'. Both Schwartz and Galanter note that the alarmist/jaundiced view is in the US often seen as a right-wing political movement.

[9]　M.J. Saks, Do We Really Know Anything about the Behavior of the Tort Litigation System – And Why Not?, [1992] *University of Pennsylvania Law Review* (UPaLRev), 1147 *et seq.*; M. Galanter, Real World Torts: An Antidote to Anecdotes, [1996] *Maryland Law Review* (MdLRev), 1093 *et seq.*

[10]　G.T. Schwartz, [2002] UIllLRev, 1076.

for Civil Justice. Although this RAND study finds indeed that only 10% of all accident victims recover any damages from the tort system (see Table 2 in the Appendix)[11], Schwartz points out that it does not provide information on the actual percentage of injuries caused by wrongful behaviour of one or more parties. This example according to Schwartz illustrates that the reassuring writings may sometimes mischaracterise the available empirical evidence just like the alarmist writings do, despite their own commitment to empirical research.[12]

He continues by discussing Carroll (1997)[13], who found some evidence for excessive claiming behaviour under tort law in a case study of routine auto accidents. Carroll came to this conclusion after having compared the pattern of claimed auto injuries in tort jurisdictions with the pattern of claimed injuries in jurisdictions that have adopted no-fault compensation programmes. Schwartz rightly states that if Carroll is correct, excessive claims are a major feature of the American tort system, but that this conclusion would be limited to the domain of auto accidents.[14]

7   However, Cooter and Ulen (2004), in a section on tort law reform in the United States, highlight that since 1991 there has been a significant *decrease* in automobile accident tort filings.[15] They ascribe this partly to declining rates for automobile accidents involving personal injury, but it is unclear how this relates to Carroll's findings. Overall, Cooter and Ulen conclude (like Galanter and Saks) that the tort-liability system is working reasonably well. E.g., discussing the domain of products liability, they note that if the vast number of asbestos claims[16] are excluded, the number of products-liability cases in the federal courts even decreased by 40% between 1985 and 1991.[17]

8   Dewees *et al.* (1996), finally, are more critical towards the tort system. The empirical evidence found by the authors evidently has convinced them that the tort system alone cannot deal with both deterrence and compensation goals. Therefore they endorse some of the shifts away from torts that took place in the United States in the 20th century, such as the introduction of no-fault compensation schemes to compensate victims of automobile accidents and the shift from tort law to workers' compensation to compensate victims of workplace accidents (see Section 3).[18]

---

[11]   D.R. Hensler *et al.* (RAND Institute for Civil Justice), *Compensation for Accidental Injuries in the United States* (1991), 107.

[12]   G.T. Schwartz, [2002] UIIlLRev, 1076–1077.

[13]   S. Carroll, *Effects of an Auto-Choice Automobile Insurance Plan on Costs and Premiums* (1997), at: http://www.rand.org.publications/CT/CT141.

[14]   G.T. Schwartz, [2002] UIIlLRev, 1077–1078.

[15]   R. Cooter/T. Ulen (*supra* fn. 3), 377.

[16]   See Section 3 for a brief discussion of asbestos cases.

[17]   R. Cooter/T. Ulen (*supra* fn. 3), 381.

[18]   D. Dewees/D. Duff/M. Trebilcock (*supra* fn. 2), 412. On p. 412 the authors write that the deterrent effect of tort, as suggested in the academic literature, is limited and uneven or cannot be established by existing empirical studies. This would suggest that 'considerable intellectual effort has been expended on models that omit some crucial facts about the real world, including high transaction costs and imperfect information'.

2.1.2 Compensation

As stated above, another (or even primary) goal of tort law would be to compensate accident victims for losses resulting from such accidents.[19] Indeed, one may wonder to what extent tort law is able to compensate victims for harm caused by others (such as reckless car drivers, polluting factories, negligent physicians and co-workers etc.) or harm caused indirectly, e.g. by having to work in an unsafe working environment. However, the important legal questions as to the determination of the causal link between the accident and the losses, and as to the amount and composition of compensation paid out via the tort system (e.g. does it also include moral damage?) are beyond the scope of this paper.

It would have been interesting to present here some kind of 'general conclusion' on the performance of the US tort system as regards compensation of personal injury victims, but this appears to be difficult. The explanation for this is straightforward: In the majority of cases by far, tort scholars have discussed specific domains of personal injury, such as medical negligence, product liability or work injury (see Section 3 for the latter). Hence general claims about the performance of the tort system as regards compensation have very rarely been made.

9

An exception is the comprehensive study by Dewees *et al.* (1996). In the final chapter of their book, the authors conclude that the tort system 'performs so poorly in compensating most victims of personal injury that we should abandon tort as a means of pursuing this compensation objective, turning instead to other instruments'.[20] The authors suggest the introduction of separate compensation systems in some accident areas, funded by risk-rated premiums. However, tort law should still play a residual role in cases involving serious harm and a clear causal link to the injurer. In that respect Dewees *et al.* refer to automobile accidents, medical malpractice and workplace accidents (torts in addition to no-fault insurance schemes) and environmental injuries (torts in addition to regulation). Product liability, the authors argue, could still be governed by tort law in the form of negligence.[21]

10

It needs to be pointed out also that the introduction of liability insurance in combination with tort law, as commonly seen in practice, might either increase or decrease the levels of compensation paid out. As Abraham (2004) puts it, liability insurance at least 'facilitates the compensation of successful plaintiffs'.[22] Although one may wonder whether the presence of liability insur-

11

---

[19] See also a long paper by J.C.P. Goldberg, Twentieth Century Tort Theory, [2003] *Georgetown Law Journal* (GeoLJ), Vol. 91, 514–583 who discusses various 'theories' about the functions of tort law, on the basis of academic debates that took place in the United States in the 20th Century.

[20] D. Dewees/D. Duff/M. Trebilcock (*supra* fn. 2), 412.

[21] D. Dewees/D. Duff/M. Trebilcock (*supra* fn. 2), Chapter 7.

[22] K.S. Abraham, Liability Insurance and Accident Prevention: The Evolution of an Idea, [2004] *Public Law and Legal Theory Working Paper Series, University of Virginia Law School*, 1.

ance would reduce safety incentives because of moral hazard[23] (i.e. whether liability insurance may undermine accident prevention), Abraham shows how in the US at first this criticism was defended in the legal literature and in legal practice 'primarily on the ground that [liability insurance] would promote the compensation of accident victims'. His paper then reviews the evolution of these ideas in the major debates in the US about insurance and tort reform, in which finally the interrelationship between liability insurance (and the loss-spreading rationale) and accident prevention were stressed.[24] Naturally, the computation of the insurance premium is an important factor also: e.g., to what extent is the premium risk-related or experience-rated?[25]

2.2 Alternatives to Tort Law

12   Alternatives to the tort system such as no-fault compensation funds and social security have been discussed extensively in the theoretical law-and-economics literature.[26] As stated above, no-fault systems and social security are not likely to have any deterrent effects unless the financial contributions to these systems are made dependent on factors relating to the accident risk or unless they are combined with some form of safety regulation. This non-deterrence argument applies in particular to the case of social security, which is generally considered as a 'Existenzsicherung': i.e. providing some basic (limited) compensation, whereas equal access to the system is usually essential (meaning there is no risk differentiation). No-fault systems have a similar characteristic in the fact that accident victims do not need to prove fault to receive compensation. However, in a no-fault system the link between the accident and the particular fund must still be proven, which may in some cases be difficult. Also, no-fault schemes generally favour particular kinds of accident victims over others who may be similarly injured by other kinds of accidents[27] (compare, e.g. falling from a ladder at home and falling from a ladder at the workplace), which cannot always easily be justified. There are, however, many variants of no-fault compensation schemes with varying degrees of residual tort liability and various types of compensation, so generalisation is impossible.[28]

---

[23] In relation to insurance, the concept 'moral hazard' simply means that the very fact that one is insured will make him or her take less than optimal care (e.g. because of bike or fire insurance), or that the demand for a certain service increases as soon as full insurance cover is available (e.g. health insurance). For the latter, see in particular K. Arrow, Uncertainty and the Welfare Economics of Medical Care, [1963] *American Economic Review* (AmEconRev), 941 *et seq.* R. Cooter/T. Ulen (*supra* fn. 3), 54 and 354–357, provides a general analysis.

[24] For more on these topics see in particular the classic law-and-economics work by Calabresi, who distinguishes between primary (victims' losses), secondary (risk spreading costs, if victims are risk-averse) and tertiary (administrative) accident costs. Liability insurance is one of the forms of loss spreading discussed by Calabresi. See G. Calabresi, *The Costs of Accidents: A Legal and Economic Analysis* (1970).

[25] I will come back to these issues below, albeit briefly.

[26] For a discussion of no-fault systems see also the contribution to book 3 by Oliphant.

[27] R.I. McEwin, No-Fault Compensation Systems, in: B. Bouckaert/G. De Geest (eds.), *Encyclopedia of Law and Economics, Volume II. Civil Law and Economics* (2000), 738.

[28] A no-fault system can be combined with *private* insurance, whereas social security itself is considered a *social* insurance. For a brief discussion of the basic principles of private insurance

As just mentioned, a combination of a no-fault or social security system and   13
safety regulation may provide a solution to this problem in the sense that then
both deterrence and compensation are, in theory, encouraged. Also, tort law
and regulation can and should be combined, according to Shavell (1984).
Shavell discusses four criteria that determine the choice between tort law and
regulation as instruments for controlling risky activities: information, insolvency risk, the threat of a liability suit and administrative costs. He concludes
that 'a complete solution to the problem of the control of risk evidently should
involve the joint use of liability and regulation, with the balance between them
reflecting the importance of the determinants'.[29] One should keep in mind,
however, that regulatory agencies may (more easily than courts) be influenced
by interest groups lobbying for regulation that serves private interests rather
than the public interest.[30]

In addition, authors tend to disagree on the actual preventive effects of the cur-   14
rent and past safety regulation in the United States. The Cato Institute (2003)
argues that safety regulation in the United States has had little (preventive) effect on the number of accidents. The authors use National Safety Council data
to show that throughout the past century accidental death rates have been decreasing in all accident domains (motor vehicle[31], work, home and public), but
add that this decline cannot be credited to regulatory agencies.[32] To the contrary, there seems to be no evidence of any additional effects of regulatory policies on this downward trend. Rather, the Cato Institute puts forward the hypothesis that improvements in societal wealth and technology have increased
our demand for safety over time: 'market forces rather than regulatory policy
have likely been the most important contributor to safety improvements since
early last century'.[33] In Section 3 I will discuss some (quite critical) empirical
papers on OSHA, the Occupational Safety and Health Administration.

Dewees *et al.* (1996) have a more moderate attitude towards (safety) regulation.   15
They conclude that government regulation has achieved modest success in the
domains of workplace safety, product liability and medical safety. However, as
regards reducing the number of environmentally-related accidents and traffic ac-

---

and social insurance, see M.G. Faure, The Applicability of the Principles of Private Insurance
to Social Health Care Insurance, Seen from a Law and Economics Perspective, [1998] *The
Geneva Papers on Risk and Insurance: Issues and Practice* (GPRIIP), 267–268. R.I. McEwin
(*supra*, fn. 27) gives an overview of law-and-economics literature on no-fault compensation
systems.

[29] S. Shavell [1984] JLS, 365.

[30] For a discussion of the private interest approach to regulation, rent seeking, and related literature: See N.J. Philipsen, *Regulation of and by Pharmacists in the Netherlands and Belgium*
(2003), 23–27. See also the contribution by Ogus to book 3.

[31] If adjusted to changes in 'driving intensity'.

[32] P. VanDoren, Occupational Safety and Health Administration, in: D. Boaz/E.H. Crane (eds.)
(Cato Institute), *Cato Handbook for Congress: Policy Recommendations for the 108th Congress* (2003) at: http://www.cato.org/pubs/handbook/hb108/hb108-35.pdf, 367.

[33] P. VanDoren (*supra* fn. 32), 368.

cidents, regulatory policies have been more successful, albeit that in some cases the costs clearly outweighed the benefits. Overall, the effectiveness of the regulatory system could be improved by improving its design and by reducing its use in areas in which it is ineffective, but also by expanding its use in areas where it is more effective (notably the control of environmental injuries).[34]

16   McEwin (2000) arrives at the following conclusion in his chapter on no-fault compensation systems in the Encyclopaedia of Law and Economics: 'Fault and no-fault compensation systems should not be considered in isolation. Fault systems can be combined with safety regulation and compulsory first- or third-party insurance systems. So can no-fault systems. From a policy perspective different combinations of insurance/safety regulation should be considered in terms of their ability to provide optimal compensation and safety as well as satisfying societal demands for 'retribution' and 'justice'. But while a considerable amount has been written, we still do not know whether no-fault insurance, taken together with other compensation sources and other incentives to take care, increases social welfare'.[35]

3. Work Injury

17   If employers and employees were able to perfectly assess the risks of industrial accidents *ex ante*, they could – according to the Coase Theorem[36] – allocate these risks in an efficient way by incorporating risk premiums in wages. In that case the employer's incentive to prevent accidents would be embodied in the wage premium. However, such an optimal allocation of risks would follow only if transaction costs are zero or negligible. In reality transaction costs in the labour market are not negligible. Dewees *et al.* (1996) state that, because of market failures in the labour market, risk premiums will not be set at the efficient level. They refer in particular to three forms of market failure: information problems regarding the risks[37], unequal bargaining power between employers and employees, and externalities[38] caused by fatal injuries.[39] These

---

[34] D. Dewees/D. Duff/M. Trebilcock (*supra* fn. 2), vi and 413–414.

[35] R.I. McEwin (*supra* fn. 27), 745.

[36] The general version of this theorem states that in the absence of transaction costs, an optimal allocation of resources (efficiency) will always follow, irrespective of the initial distribution of property rights (irrespective of the prevailing liability rule). It is a central theorem in the law and economics literature and is based on R. Coase, The Problem of Social Cost, [1960] *Journal of Law and Economics* (JLE), 293 *et seq.* For a general description, I refer to R. Cooter/T. Ulen (*supra* fn. 3), 85–96, and R.A. Posner (*supra* fn. 3), 7 and 49–52.

[37] On the modelling of underestimation of risks by workers: S.A. Rea, Workmen's Compensation and Occupational Safety under Imperfect Information, [1981] *American Economic Review* (AmEconRev), 80 *et seq.*

[38] Externalities appear in the wage contract setting because of the existence of social costs in excess of the private costs that fall on the parties to the bargain. These social costs are not taken into account (i.e., will not be allocated) in the private bargain between the (representatives of) employers and employees. For a general description of the concept 'externality', I again refer to R. Cooter/T. Ulen (*supra* fn. 3), 44–46, and R.A. Posner (*supra* fn. 3), 71.

[39] D. Dewees/D. Duff/M. Trebilcock (*supra* fn. 2), 347–348. The authors also discuss some empirical literature on risk premiums here.

are, indeed, exactly those situations where transaction costs in the labour market are high.[40]

Hence, in order to give incentives to employers and employees to prevent industrial accidents and occupational diseases, and in order to compensate the victims of such accidents and diseases when they do occur, some kind of intervention in the labour market is required, e.g. in the form of tort liability, safety regulation, or a no-fault system. In practice, each country uses its own combination of these and other instruments, as we show also in the other contributions to this book, which deal with the compensation systems in four European countries.[41]

In this section I will review the existing empirical literature on the shift from   18
tort liability to workers' compensation in the United States, as well as the literature that analyses the workers' compensation system itself. First, however, some necessary background information on the American system and the actual number of industrial accidents and occupational diseases will be provided (3.1). The deterrent effects of American tort law and workers' compensation will be discussed in Section 3.2, followed by a discussion of the compensation effects (3.3). I will also address, albeit briefly, the safety regulation enforced by the Occupational Health and Safety Administration (OSHA). The discussion that follows is not meant to be complete. Because the literature on the workers' compensation system is vast[42], I will select only some of the landmark papers and books.

3.1 Background Information

According to estimations by the National Safety Council, there were 4,500   19
work-related 'unintentional-injury deaths' in the United States in 2003. This implies a rate of 3.2 deaths per 100,000 workers. In addition there were 3,400,000 'disabling injuries', giving rise to 70 million workdays lost in that same year (plus an estimated 55 million days lost in future years). The total economic costs of the occupational deaths and injuries in 2003 are estimated at $156.2 billion. These costs include – among other things – wage and productivity losses of $78.3 billion, medical costs of $30.9 billion, and administrative expenses of $28.7 billion. Table 1 presents the number of deaths, the number of workers and the death rates for the period 1992–2003. The numbers suggest that in recent years there has been a decrease in the death rate.[43]

---

[40] For more information on the concepts discussed in this paragraph, see N.J. Philipsen (*supra* fn. 30), Chapter 2.

[41] *Supra*, note 1.

[42] However, much literature that was available in the library of the University of Illinois appears to be hardly available in Europe or on the internet.

[43] National Safety Council, *Injury Facts: 2004 edition* (2004), 48–51. This report also contains information by industry and by state, as well as some estimates of the prevalence of occupational diseases. See pp. 47–84 and 159–161.

Table 1 Unintentional Work-Injury Deaths and Death Rates, U.S., 1992–2003[44]

Year	Deaths	Workers	Death rate
1992	4,965	119,168,000	4.2
1993	5,034	120,778,000	4.2
1994	5,338	124,470,000	4.3
1995	5,015	126,248,000	4.0
1996	5,069	127,997,000	4.0
1997	5,160	130,810,000	3.9
1998	5,117	132,772,000	3.9
1999	5,184	134,688,000	3.8
2000	5,022	136,402,000	3.7
2001	5,042	136,246,000	3.7
2002	4,716	137,731,000	3.4
2003	4,500	138,988,000	3.2

20   With regard to 'occupational illness', Bureau of Labor Statistics data state that approximately 294,500 cases were recognized or diagnosed by employers in 2002. The incidence rate per 10,000 full-time workers was 33.3. The most common illness in the United States in 2002 were 'skin diseases or disorders' with 44,900 new cases. Time-series of data on occupational diseases are unfortunately not given in the report and are hard to find.[45] The Bureau of Labor Statistics website does, however, provide data for recent years[46], and a Cato Institute report from 1997 includes a figure showing no clear trend in the 'nonfatal workplace injury and illness rate' in the period 1973–1993.[47] In 1973 firms were for the first time required to report industrial accidents and diseases. We should remember, however, that there may be a huge underreporting of cases of (in particular) occupational diseases, a problem common in most countries.[48]

21   With respect to workplace fatalities and (to some extent) injuries, older data are somewhat easier to find, although at least until the early 1990s estimates of the number of workplace fatalities and injuries in the United States varied widely between sources. Dewees, Duff and Trebilcock (1996) mention estimates for the number of workplace fatalities around the year 1989 ranging from 3,000 to 11,000 per year.[49] Going back in time much further, the Nation-

---

[44] National Safety Council (*supra* fn. 43), 49. Numbers for 2003 are preliminary. Death rate: deaths per 100,000 workers.

[45] National Safety Council (*supra* fn. 43), 82.

[46] See: http://www.bls.gov/iif/.

[47] T.J. Kniesner/J.D. Leeth, Occupational Safety and Health Administration, in: D. Boaz/E.H. Crane (eds.) (Cato Institute), *Cato Handbook for Congress: 105th Congress* (1997), at: http://www.cato.org/pubs/handbook/hb105-36.html.), 2–3. The original source of the data presented there is U.S. Department of Labor, OSHA.

[48] See the other contribution by Philipsen to this book.

[49] See also D. Dewees/D. Duff/M. Trebilcock (*supra* fn. 2), 346.

al Safety Council argues that the death rate decreased by 93% between 1912 and 2003: In 1912 an estimated 18,000 to 21,000 workers lost their lives at work in a work force that was only one fourth of the size of today's work force and producing one ninth of today's goods and services.[50] A graph provided by the Cato Institute shows that this decrease had been gradual, at least from 1933 onwards, with only occasional temporary increases in the death rate.[51] However, concrete data for the first half of the 20th century are difficult to find, if they exist at all. Little, Eaton and Smith (2004) note that in the 1960s there was a substantial increase in injury rates, although there are no concrete figures to back this up. By the end of that decade 'approximately 14,000 people were killed on the job and millions were injured each year'.[52] The high estimated numbers of injuries in that period eventually led to the introduction of the Occupational Safety and Health Act in 1970 and the establishment of the Occupational Safety and Health Administration (OSHA), the federal agency that sets and enforces workplace standards in the United States.[53] In the period 1972–1980, the frequency and average duration of the more serious and expensive injury cases increased, whereas the overall injury rate fell.[54] The National Safety Council (2004) presents a graph with occupational incidence rates (injuries and illnesses per 100 employees) for the period 1982–2002, based on data provided by the Bureau of Labor Statistics. This graph shows a stabilisation of the occupational incidence rate in the 1980s (with some minor increases in the second half of the decade) and a gradual decrease since 1992.[55]

With respect to the compensation of victims of industrial accidents and occupational diseases in the United States, tort actions have largely been replaced by no-fault compensation schemes, whereas the deterrence function of tort has to some extent also been replaced by occupational health and safety regulation.[56] The introduction of no-fault schemes took place mostly in the early twentieth century. While in 1900, negligence[57] was used as a basis for deter-   22

---

[50] National Safety Council (*supra* fn. 43), 48.

[51] T.J. Kniesner/J.D. Leeth (*supra* fn. 47), 2. The authors use own calculations based on National Safety Council data.

[52] J.W. Little/T.A. Eaton/G.R. Smith, *Workers' Compensation: Cases and Materials* (5th edn. 2004), 43.

[53] For an evaluation of OSHA in its first years, see in particular W.Y. Oi, On the Economics of Industrial Safety, [1973–1974] *Law and Contemporary Problems* (LCP), 694 *et seq.*

[54] J.D. Worrall/D. Appel, Some Benefit Issues in Workers' Compensation, in: J.D. Worrall/D. Appel (eds.), *Workers' Compensation Benefits: Adequacy, Equity and Efficiency* (1985), 7–8.

[55] National Safety Council (*supra* fn. 43), 60.

[56] D. Dewees/D. Duff/M. Trebilcock (*supra* fn. 2), 346.

[57] At the time, three major defenses existed to the benefit of the employer: negligence of fellow-servants, voluntary assumption of risk and contributory negligence by injured workers. Some states relied on the common law, while in others the nature of the negligence rules was controlled by statute. See, e.g., J.R. Chelius, Liability for Industrial Accidents: A Comparison of Negligence and Strict Liability Systems, [1976] *Journal of Legal Studies* (JLS), 298–301; P.V. Fishback, Liability Rules and Accident Prevention in the Workplace: Empirical Evidence from the Early Twentieth Century, [1987] *Journal of Legal Studies* (JLS), 307; D. Dewees/D. Duff/M. Trebilcock (*supra* fn. 2), 349–350; and J.W. Little/T.A. Eaton/G.R. Smith (*supra* fn. 52), 6–20.

mining liability for industrial accident costs in all states of the US, things
started to change in 1911. In that year, the first states started switching to a
system of 'shared strict liability' known as *workers' compensation*. By 1949
all states had switched to this system.[58] Workers' compensation functions as a
no-fault insurance system. In case of an accident, employers pay a govern-
mentally determined amount to the victim-employees or their heirs, irrespective of the cause of the accident. Of course, there is one important criterion:
The accident must be work-related.[59] Employees covered by workers' compensation are barred from any negligence proceeding against their employer.[60]
There have been, however, many American cases dealing with workplace-re-
lated product liability claims (among them many asbestos claims), because
these are not excluded against product manufacturers by the workers' compensation law.[61]

23   The shift from tort law to workers' compensation took place mainly because
the common-law remedies for injured employees came to be widely regarded
as providing too few incentives for safety at the workplace, while being un-
fairly biased toward the interests of employers.[62] The new system emerged
with a *quid pro quo*. In exchange for giving up their right to tort actions, employees receive swift and certain payment, without having to demonstrate the
employer was at fault. And in exchange for this, employers enjoy limited liability for industrial accidents and occupational diseases. As far as benefits are
concerned, workers' compensation requires employers both to provide indemnity benefits[63] and to reimburse medical costs. In order to do so, employers
have to purchase insurance, either from a private insurance carrier, a state insurance fund or by self-insuring.[64] The premium depends *inter alia* on the size
of the employer, experience (accident record of the firm), classification (type
of industry) and the insurance arrangement, and is paid by the employer as a
percentage of total payroll.[65]

---

[58] J.R. Chelius, [1976] JLS, 298. Note that the workers' compensation laws are state laws. See
also L. Darling-Hammond/T.J. Kniesner (RAND Institute for Civil Justice), *The Law and Economics of Workers' Compensation* (1980), 7–10; and J.D. Worrall/D. Appel (*supra* fn. 54), 3.

[59] The workers' compensation system applies to injuries and diseases 'arising out of and in the
course of employment'. J.D. Worrall/D. Appel (*supra* fn. 54), 1.

[60] J.R. Chelius, [1976] JLS, 300.

[61] D. Dewees/D. Duff/M. Trebilcock (*supra* fn. 2), 346; M.J. Moore/W.K. Viscusi, *Compensation
Mechanisms for Job Risks: Wages, Workers' Compensation, and Product Liability* (1990), 10.
For a statistical analysis of 1,447 of such claims, see pp. 136–161 of the latter. See also P.M.
Danzon, Compensation for Occupational Disease: Evaluating the Options, [1987] *Journal of
Risk and Insurance* (JR&I), 263–264 and 276–277.

[62] L. Darling-Hammond/T.J. Kniesner (*supra* fn. 58), 7.

[63] The different categories of indemnity benefits (or: cash benefits) and the level of benefits will
be discussed in Section 3.2.

[64] There are differences among the states, e.g. some have state funds competing with private
insurers, while others have exclusive state funds.

[65] This paragraph is based on J.D. Worrall/D. Appel (*supra* fn. 54), 3–5; and D. Dewees/D. Duff/
M. Trebilcock (*supra* fn. 2), 379–381; 387–391. My description here is obviously very brief.
For more information I refer to the two studies just mentioned.

3.2 Deterrence

The law-and-economics literature is clear in defining the goal of a liability 24
system for industrial accidents: Such a system should minimise the sum of
(expected) accident costs and prevention costs.[66] The former category includes
not only the costs of lost wages and medical expenses, but also the costs of
production losses and 'pain and suffering'. The latter category includes expenses made by the employer to prevent accidents (such as the costs of guard-
ing machines or slowing down production) and the administrative costs of the
liability system. The liability system should be designed in such a way that it
gives incentives to employers and employees to prevent those accidents for
which prevention costs are lower than accident costs.[67] In other words, it
should deter those accidents. I already discussed the more general 'economic
theory of tort law'[68], and especially the debate surrounding the deterrent effects of tort law, in Section 2.1. Now I will look at the actual deterrence of tort
law and workers' compensation in the domain of work injury in the United
States.[69]

Dewees *et al.* (1996) state that, at least until 1996, there had been very little 25
empirical study of the deterrence effect of tort liability for occupational injury.[70] Two American studies by, respectively, Chelius (1976) and Fishback
(1987), concentrate on the early twentieth century and analyse the effect of the
change from a tort regime to a workers' compensation regime on accident
rates. They do, however, reach opposite conclusions. Chelius, who made use
of data on non-motor vehicle machinery fatalities from 1900–1940, concluded
that the death rate decreased *more* in jurisdictions where workers' compensation was introduced than in jurisdictions where tort liability had been expanded (although in both types of cases a decrease was found).[71] Fishback, who
analysed data on fatal coal mining accidents from 1903–1930, found that the
shift from very restricted tort liability to either workers' compensation or expanded tort liability led to higher accident rates.[72] This rather surprising result
may according to Fishback be explained by the fact that the cost of supervision of workers (safety) in the mining industry was very high, while the abolition of the fellow-servant rule[73] reduced workers' incentives to look after the
safety of their co-workers. In factories with dangerous machinery, such as

---

[66] P.M. Danzon, [1987] JR&I, 264, argues that the social costs associated with accidents have
four sources: prevention costs; the costs of compensating injuries; litigation, enforcement and
other overhead costs; and the disutility of uninsured risks.

[67] J.R. Chelius, [1976] JLS, 294; P.V. Fishback, [1987] JLS, 306. See also W.Y. Oi, [1973–1974]
LCP, 669–680.

[68] *Supra*, note 3 and accompanying text.

[69] R.G. Ehrenberg, Workers' Compensation, Wages, and the Risk of Injury, in: J.F. Burton (ed.),
New Perspectives in Workers' Compensation (1988), 74–78, presents a short economic analysis
of the potential effects of workers' compensation systems on injury rates, the number of
claims, the duration of claims and the magnitude of compensating wage differentials.

[70] D. Dewees/D. Duff/M. Trebilcock (*supra* fn. 2), 352.

[71] J.R. Chelius, [1976] JLS, 306.

[72] P.V. Fishback, [1987] JLS, 322.

[73] *Supra* fn. 57.

those studied by Chelius, supervision costs for the management or foremen are probably lower, which might explain the very different results found by both authors. This ultimately leads to the conclusion that the least cost avoider (management, foremen, co-workers) may vary from industry to industry. This would imply also that the optimal liability rule may vary from industry to industry and changes with technology.[74]

26   Various authors have studied the deterrent effects of experience rating in the workers' compensation system. In that respect, Chelius and Smith (1983) attempted to gain some insight on the degree to which employer injury prevention activities are affected by the insurance arrangements used in workers' compensation. Surprisingly, they found that experience rating has no measurable effect on employer safety, although they note that they had to use a rather crude measure of marginal premium cost. Their result contrasted with some previous findings by authors such as Victor and Butler.[75] Dewees *et al.* (1996) mention two empirical studies in particular, which both conclude that experience rating under workers' compensation did have some significant effects on the actual accident rates in the 1970s.[76] Moore and Viscusi (1991), analysing fatalities under the workers' compensation system in general, come to the most surprising conclusion. They found that workers' compensation is a 'driving force in reducing fatalities at the workplace': Without workers' compensation (and without tort!), industrial fatality risks could have risen by more than 40%, they say. This means, according to the authors, that it has saved the lives of almost 2,000 workers per year.[77] It would also imply that workers' compensation has had much more deterrent effects than OSHA safety regulations, for which the reduction in risk levels has been estimated between 2 and 4% only.

And although this does not concern the United States, I should also mention a German study by Kötz and Schäfer (1993) on the sugar industry. Namely, the authors conclude that there were far fewer accidents after a risk-rated premium was introduced in the German workplace compensation scheme than before. Apparently the economic incentives generated by the risk-rated premium induced managers of firms in the German sugar industry to take measures of accident prevention.[78]

27   Overall, the existing empirical evidence sometimes reaches different conclusions. Or, as Dewees *et al.* (1996) put it: 'the evidence does not establish that tort liability will reduce workplace injuries: It indicates that it might reduce those injuries in some industries in the absence of a regulatory regime, al-

---

[74] P.V. Fishback, [1987] JLS, 325; D. Dewees/D. Duff/M. Trebilcock (*supra* fn. 2), 353.

[75] J.R. Chelius/R.S. Smith, Experience-Rating and Injury Prevention, in: J.D. Worrall (ed.), *Safety and the Work Force* (1983), 130.

[76] These are studies by Ruser (1985) and Worrall and Butler (1985). For more details, see D. Dewees/D. Duff/M. Trebilcock (*supra* fn. 2), 381.

[77] M.J. Moore/W.K. Viscusi (*supra* fn. 61), 9.

[78] H. Kötz/H. Schäfer, Economic Incentives to Accident Prevention: An Empirical Study of the German Sugar Industry, [1993] *International Review of Law and Economics* (Int'lRevL&Econ), 19 *et seq.*

though in those circumstances workers' compensation appears to have a greater deterrent effect'.[79] One possible explanation for the tort system not having strong downward effects on the number of workplace injuries could be the fear that workers (wary of unemployment) may have of creating an adversarial relationship with their employer, which may deter them from testifying against their employer.[80] This would be less of a problem under workers' compensation. The evidence regarding the deterrent effect of workers' compensation is mixed, but overall slightly positive.[81]

While there is apparently some doubt about the actual deterrent effect of tort law and workers' compensation in cases of industrial accidents, there seems to be more consensus about the deterrent effect in cases of occupational disease. That is, this effect is probably very small, for various reasons. In cases of occupational disease there are often latency problems.[82] Many times the link between an occupational hazard and a certain disease is only discovered years later – if at all – and sometimes only upon the death of the victim. In a Canadian study, Dewees (1986) found that, for asbestosis, the long latency period 'greatly reduces the present value of death and disease claims under the high corporate discount rate compared to the immediate and short-term costs of control'.[83] In addition, employees themselves often fail to know that their diseases are work-related, insurers frequently contest disease claims, and some specific diseases may not be compensated by the workers' compensation system. Generally, therefore, many authors are very sceptical towards the deterrent effects of tort law and the current workers' compensation systems in the area of occupational diseases.[84] Even with respect to the asbestos litigation in the United States, the most massive product liability[85] litigation ever experienced there, the high amount of cases may not have achieved ends substantially different from those achieved by other means such as the safety regulations

28

---

[79] D. Dewees/D. Duff/M. Trebilcock (*supra* fn. 2), 355.

[80] See also J.T.A. Gabel, Escalating Inefficiency in Workers' Compensation Systems: Is Federal Reform the Answer?, [2000] *Workers' Compensation Law Review* (Workers'CompLRev), 74.

[81] D. Dewees/D. Duff/M. Trebilcock (*supra* fn. 2), 381–382.

[82] See, e.g., P.M. Danzon, [1987] JR&I, 270–272. In her paper, Danzon evaluates alternative rules of liability and compensation for occupational disease in the United States. The options considered are employer liability under workers' compensation, tort liability of product manufacturers, first party insurance by employees, and potential government programs such as funds.

[83] D.N. Dewees, Economic Incentives for Controlling Industrial Disease: The Asbestos Case, [1986] *Journal of Legal Studies* (JLS), 318.

[84] See for further references D.N. Dewees, [1986] JLS, 290. In addition, Dewees shows in his asbestosis study that even full experience rating of workers' compensation assessments, whereby premiums rise with claims costs, did not create substantial incentives for the asbestos-cement pipe plant to reduce worker risks (p. 317). An elaborate discussion of the problems of dealing with occupational diseases under workers' compensation can be found in J.R. Chelius, The Status and Direction of Workers' Compensation: An Introduction to Current Issues, in: J.R. Chelius (ed.), *Current Issues in Workers' Compensation* (1986), 10–13, and the accompanying papers.

[85] Since the 1973 *Borel* decision, American workers have been allowed to sue suppliers of hazardous products (for failure to warn of the hazard) instead of their employers. This led to a huge amount of asbestos cases in the 1970s and 1980s.

of the 1970s and 1980s that limited workers' exposure to asbestos. At least, that is the conclusion of a brief literature review by Dewees *et al.* (1996).[86]

29   Enforcement of the occupational safety and health standards is very much centralized in the United States. That is, the enforcement of these standards should be guaranteed by OSHA through random inspections of workplaces along with targeted inspections in exceptional cases. Also, employees have the right to file complaints with OSHA, but they rarely do this because of lack of information and threats of employer reprisal. Another – and much bigger – problem is that OSHA's enforcement capabilities have always been low, and today it even has fewer staff than it did in the 1980s, while the number of workplaces has grown.[87] In a recent paper evaluating the practice of workplace safety regulation in the United States, Klaff (2005) states that 'as is the case with a large variety of legislation, the statutory intent and provisions of the [OSH] Act do not necessarily correspond directly with the day-to-day reality in the workplace. One of the most basic reasons for this is the failure of Congress to back-up its ambitious enforcement regime with adequate funding'.[88] In a report published by the Cato Institute in 1997, which even suggests to shut down OSHA altogether, it is argued that the post-1970 drop in workplace deaths[89] cannot be credited to OSHA. In fact, the downward trend began well before its creation. The authors claim that 'the vast majority of studies has found no statistically significant reduction in the rate of workplace fatalities or injuries due to OSHA'.[90] In addition, the costs of OSHA are estimated to be much higher than its benefits, according to the Cato Institute (referring to various researches in the 1990s).[91] The authors conclude that deterrence of workplace accidents comes mostly from the workers' compensation insurance system[92] and risk premiums in wages; increasing OSHA's resources would not change much, given its current ineffectiveness in standard setting, inspections and fines.[93]

3.3 Compensation

30   Let us look first at the compensation of accident victims under the tort system in the early twentieth century. Referring to the mining industry, Fishback (1987) argues that the amount of compensation received by accident victims was less than the full amount of accident costs under each liability rule – be it negligence, modified negligence or shared strict liability. However, to some

---

[86] D. Dewees/D. Duff/M. Trebilcock (*supra* fn. 2), 354–355.
[87] D.B. Klaff, Evaluating Work: Enforcing Occupational Safety and Health Standards in the United States, Canada and Sweden, [2005] *University of Pennsylvania Journal of Labor and Employment Law* (UPaJLab&EmpL), 613.
[88] D.B. Klaff, [2005] UPaJLab&EmpL, 628.
[89] Supra, section 3.1.
[90] T.J. Kniesner/J.D. Leeth (*supra* fn. 47), 3.
[91] T.J. Kniesner/J.D. Leeth (*supra* fn. 47), 3–4.
[92] Cf. the findings of M.J. Moore and W.K. Viscusi (*supra* fn. 61), presented [in margin number 26].
[93] T.J. Kniesner/J.D. Leeth (*supra* fn. 47), 5–6.

extent this compensation was supplemented by risk premiums in wages and by insurance through miner's relief funds.[94] Dewees *et al.* (1996) stress that plaintiffs in that period had to bear large litigation costs and that the majority of cases in that period were settled out of court, with payments far below full compensation for pecuniary losses. Because many employers had 'deep pockets' it was difficult and costly for victim-employees to negotiate with them. The authors hence conclude their review of the performance of the tort system by stating that there was a significant undercompensation for losses.[95] As to the percentage of accident victims actually receiving compensation from employers, estimates range from 6 to 30%.[96] Altogether, the empirical record of the tort system as far as compensation is concerned, is meagre.[97]

Under the *workers' compensation system*, compensation for the victim consists of reimbursement for medical costs as well as indemnity benefits.[98] Compensation for pain and suffering is not available. There are four categories of indemnity benefits, depending on the disability type of the victim: temporary total disability, permanent total disability, permanent partial disability and benefits following on a fatal accident (burial costs and benefits paid to the family of the deceased). The largest category in terms of total indemnity costs is permanent partial disability (pensions paid to the worker to compensate for functional limitations or partial earning loss for life). Although these four categories exist in all the American states, the actual level of benefits differs among the states. There are also differences in waiting periods and time limits for compensation.[99]

31

Generally, indemnity payments provide the victim with two-thirds of gross wages for temporary or permanent disability, up to a certain maximum determined by the state. This implies that the *real* rate of wage replacement can be lower than 66.6% for high wage workers, or higher than 100% for low wage workers. Often a 'schedule approach' is used, in which first the percentage for a particular disability is specified (e.g., 35% for loss of leg above the knee), and then this percentage is multiplied by the lost wage replacement figure to determine the amount of the benefit.[100] For some injuries, states may fix the benefit at a certain amount, e.g. $5,000 for a lost arm, meaning that the

---

[94] P.V. Fishback, [1987] JLS, 324.
[95] D. Dewees/D. Duff/M. Trebilcock (*supra* fn. 2), 360.
[96] J.R. Chelius, [1976] JLS, 300 (based on reports from the early twentieth century).
[97] In this respect Abraham, [2004] Public Law and Legal Theory Working Paper Series, 20–21, refers to a 1910 study by Eastman, which focused on the deaths of married workers as a result of industrial accidents in Pittsburgh. The Eastman study demonstrated the huge gap between losses suffered by workers and the amounts recovered in tort, when they recovered at all. Abraham, however, uses this example also to show that in the early 20th century people were much more concerned with compensation issues than with accident prevention.
[98] The National Academy of Social Insurance estimated the total workers' compensation payments for the year 2001 at $49.4 billion, of which $22.0 billion 'medical and hospital' and $27.4 billion 'wages'. J.W. Little/T.A. Eaton/G.R. Smith (*supra* fn. 52), 2.
[99] J.D. Worrall/D. Appel (*supra* fn. 54), 4–5; D. Dewees/D. Duff/M. Trebilcock (*supra* fn. 2), 390.
[100] J.D. Worrall/D. Appel (*supra* fn. 54), 5–6; D. Dewees/D. Duff/M. Trebilcock (*supra* fn. 2), 390–391.

benefit does not depend on the victim's actual job or on the actual losses.[101] Some states use a 'two-part award', to compensate workers for both the disability itself and to replace lost wages.[102] Finally, it should be mentioned that indemnity benefit recipients may still receive additional forms of insurance benefits, such as social security or private disability payments.[103]

32   There has been some criticism as to the computation of the benefits. One of the reasons for this is that in many states long-term awards (permanent and death benefits) have never been indexed to inflation, which has led to a continuous decrease in the real value of the award received. Another reason for criticism is the fact that benefits tend to ignore the worker's age and career stage. There has also been criticism as to the compensation for occupational diseases, because awards for occupational diseases are generally much lower than the awards paid out in cases of industrial accidents. Moreover, Little, Eaton and Smith (2004) note that only about 5% of all occupational disease victims receive compensation.[104] And in his study about asbestosis, Dewees (1986) found that workers' compensation awards in the case of premature death fall far short of full compensation valued by either the workers' willingness to pay or social willingness to pay.[105]

33   It needs to be mentioned, however, that at least until the early 1980s very little was known about the degree to which workers' compensation actually protected workers from economic hardships. There were estimations that in some situations the real protection was less than the two-thirds of income that most states were seeking to replace, but it is difficult to check such estimates as even 'the subtle calculations required to estimate income lost through permanent, partial and temporary total disabilities [did not] appear in the technical literature dealing with workers' compensation'.[106] After recommendations by the so-called National Commission on State Workmen's Compensation Laws in 1972, benefit levels increased and coverage extended in almost all of the states in the following years. Moore and Viscusi (1990) show that around 1977 benefit levels were still suboptimal, but that dramatic increases in workers' compensation benefit levels in the late 1970s and early 1980s 'closed the benefit inadequacy gap'.[107] This also led to higher premiums for the employer,

---

[101] R. Cooter/T. Ulen (*supra* fn. 3), 386.
[102] D. Dewees/D. Duff/M. Trebilcock (*supra* fn. 2), 391. The compensation for the disability itself is called 'impairment benefit' and the replacement for lost wages is called 'disability benefit'; see L. Darling-Hammond/T.J. Kniesner (*supra* fn. 58), 20–23.
[103] R.G. Ehrenberg (*supra* fn. 69), 74. According to J.W. Little/T.A. Eaton/G.R. Smith (*supra* fn. 52), 70, 'in the US the interrelationship between workers' compensation systems, which are largely privately financed, and the ever-expanding social security system has become complex'.
[104] J.W. Little/T.A. Eaton/G.R. Smith (*supra* fn. 52), 277. And of all benefits paid under the workers' compensation system, only 2–3% are for occupational disease claims.
[105] This, according to Dewees, is also the main reason why the optimal control (safety) level as enforced by the firm is lower than that for society. See D.N. Dewees, [1986] JLS, 317.
[106] L. Darling-Hammond/T.J. Kniesner (*supra* fn. 58), xviii.
[107] M.J. Moore/W.K. Viscusi (*supra* fn. 61), 5.

which was viewed with alarm by some. Others, among them Moore and Viscusi themselves, argue that the workers' compensation system pays for itself, because in assessing the net cost of indemnity benefits to firms, one must include the wage offset resulting from the benefits. In other words, the net cost to the firm is defined as the total increase in premiums minus wage reductions (which are the result of workers demanding a lower wage premium).[108]

One may wonder if the increase in workers' compensation benefits has had [34] any effect on the number and duration of claims in the United States. Ehrenberg (1988) argues that such an increase is likely to lead to moral hazard[109] by workers, because they are less motivated to take precautions (i.e., they may be less cautious at the workplace) and have a higher incentive to file claims for a given accident level. Also, higher benefits may lengthen the duration of the recovery period.[110] Indeed, Worrall and Appel (1985) report that in the early and mid 1980s there was a growing body of empirical evidence suggesting a positive association between benefits and indemnity claims or injury report filing.[111] The RAND Institute for Civil Justice reported in 1980 that the increased benefits had led to greatly increased numbers of lost time claims and lengthened duration of benefits, according to studies conducted by the insurance industry.[112]

More generally, there is also evidence that 'the substantial changes of the [35] 1970s in workers' compensation coverage and benefits, together with increased system usage by workers, resulted in dramatic increases in employer costs'.[113] Danzon and Harrington (1998) mention in that respect that benefit costs per $100 dollar of payroll increased from $0.95 in 1978 to $1.56 in 1989: in the period 1978–1984 the average annual rate of increase was 4.2% and in the period 1984–1989 it was 6.2%. They mention also that the average indemnity and medical cost per case involving lost time from work increased rapidly in the 1980s before leveling off in the 1990s.[114] As a result of the rising costs of workers' compensation insurance, employers started lobbying and many states imposed price controls ('rate suppression'). Using empirical state-level data, Danzon and Harrington argue that – contrary to its initial intent – rate suppression leads to higher costs for the workers' compensation system as a whole and ultimately to higher premiums (and probably even to more workplace injuries).[115] The explanation for this surprising result is that price controls reduce supply in the regular insurance market, which leads to

---

[108] M.J. Moore/W.K. Viscusi (*supra* fn. 61), 68 and 134; R.G. Ehrenberg (*supra* fn. 69), 75. Naturally, employers may also pass on (parts of) the costs of workers' compensation in the product prices. See, e.g., J.W. Little/T.A. Eaton/G.R. Smith (*supra* fn. 52), 67.

[109] *Supra*, note 23.

[110] R.G. Ehrenberg (*supra* fn. 69), 76–77.

[111] J.D. Worrall/D. Appel (*supra* fn. 54), 12–13.

[112] L. Darling-Hammond/T.J. Kniesner (*supra* fn. 58), 22.

[113] J.R. Chelius (*supra* fn. 84), 2.

[114] P.M. Danzon/S.E. Harrington, Rate Regulation of Workers' Compensation Insurance: How Price Controls Increase Costs (1998), 2.

[115] P.M. Danzon/S.E. Harrington (*supra* fn. 114), 107.

larger residual insurance markets, which in turn leads to a growth in claim costs.[116]

4.   Concluding Remarks

36   In this paper I presented an overview of existing empirical literature on the workers' compensation system in the United States. Clearly, many authors have addressed the question to what extent workers' compensation can provide for deterrence and/or compensation of work injury (i.e. industrial accidents and occupational diseases). On a more general level, there has also been an ongoing discussion in the United States (and elsewhere) on the objectives of tort law and on the role of other 'compensation systems' such as social insurance and no-fault systems. Again, questions of deterrence and compensation have been central in this discussion.

37   I discussed some of the latter, more general literature (i.e. not dealing with work injury specifically), in Section 2. First a distinction was made between the two opposing views of tort law: the so-called 'alarmist view' and the 'reassuring view'. Although it seems that, at least to some extent, adherents of the 'alarmist view' have misrepresented empirical evidence in order to show that there is a claims culture, adherents of the 'reassuring view' (who reject the claim that there is a claims culture) may themselves have drawn conclusions too quickly in some cases. Nevertheless, the empirical evidence presented in Section 2.1 – although at times contradicting – showed us that in the United States much information on the functioning of the tort system is available. Moreover, it showed us that much of this information has already been reviewed by American scholars in long and interesting papers. Section 2.2 dealt briefly with alternatives to the tort system as discussed in the law-and-economics literature. I also introduced the reader to the lively debate on the effectiveness of safety regulation in the United States. Various authors have suggested that a combination of different instruments is required in order to (attempt to) achieve both optimal compensation for and deterrence of accidents. Moreover, Dewees *et al.* (1996) in that respect seem to suggest that each category of accidents (be it work-related, environmental, product-related, medical or traffic-related) warrants a different combination of instruments and hence a larger or smaller (additional) role of tort law.

38   The specific domain of work injury was central in Section 3. After having presented some statistics on industrial accidents and occupational diseases in the United States, I gave a short historical overview of the shift from tort law to workers' compensation in the early twentieth century (Section 3.1). As already stated above, there is a huge and highly interesting literature on workers' compensation in the United States, some of which is hardly available in Europe. Among this literature are many empirical papers dealing with the effects (on deterrence and compensation) of the shift from tort law to workers' compen-

---

[116]   P.M. Danzon/S.E. Harrington (*supra* fn. 114), 81.

sation systems, the impact of OSHA (safety regulation) on deterrence of accidents, the deterrent effect of risk-rated premiums, the relationship between benefit levels and employer costs, and related themes. Naturally, I refer to Sections 3.2 and 3.3 for details.

Although it is difficult to draw overall conclusions, as the various papers presented us with mixed and sometimes contradicting results, one could make the following observations. The effects of the introduction of the workers' compensation system seem to have been *slightly positive* as regards the deterrence of accidents and the (swift and reasonable) compensation of victims of industrial accidents. However, it should be noted that this does *not* apply to occupational diseases. Also, it must be noted that it appears to be difficult to exactly measure the performance of the workers' compensation system as far as compensation is concerned, because workers' compensation, social insurance and private insurance are often intertwined. An additional observation is that the impact of OSHA on deterrence seems to have been regarded as *low* by most authors, mainly because of enforcement problems. The experience rating in the workers' compensation system has generally been found to be more effective.

Notwithstanding the fact that hard conclusions cannot be drawn, Little, Eaton   39
and Smith (2004) are sure that workers' compensation is here to stay, as it 'is now a stable part of the legal and economic employment enterprise throughout the United States. Although the details of the laws are being constantly adjusted, no proposal to repeal them or to replace them with drastically different plans is likely to receive generous attention anywhere'.[117] There has also been some criticism, e.g. expressed by Gabel (2000), who argued that workers compensation has 'evolved away from its original goals of efficiency, predictability and fairness to a current state of disarray amidst inconsistent state case law and federal regulation'.[118] The author argues that currently there are conflicts between the state-administered workers' compensation systems and federal regulation, which could be solved by nationalising workers' compensation. Clearly, as in Europe, the debate about the pros and cons of the US workers' compensation system is far from over.

Appendix

Table 2 gives an overview of the various compensation sources for work-related accidents and other kinds of accidents in the United States. These data have been provided by the RAND Institute for Civil Justice and concern averages per year for the period around 1991 (see also above, Section 2.1.1). According to these numbers, in that period 60% of all work-related accident victims received workers' compensation; almost one third of all victims received money

---

[117] J.W. Little/T.A. Eaton/G.R. Smith (*supra* fn. 52), 71.
[118] J.T.A. Gabel, [2000] Workers'CompLRev, 71. The federal legislation discussed in this respect by Gabel consists of the Americans with Disabilities Act (1994) and the Family and Medical Leave Act (1994 and 1997).

from their own health insurance and only 7.5% received compensation via the tort liability system.

Table 2 Persons Compensated for Loss and Sources of Reimbursement[119]

	All accidents	Work-related	Motor vehicle	Other accidents
Persons receiving comp. (millions)	23.4	7.5	3.9	12.0
Total compensation (billions of dollars)	$ 109.4	$ 45.0	$ 25.7	$ 38.7
Sources of compensation (percent of those receiving compensation)				
Injured person's own insurance				
– Health insurance	59.1%	32.7%	42.6%	76.4%
– Auto, accident, other	9.9%	6.2%	25.0%	6.9%
Workers' compensation	14.9%	59.7%	0.0%	0.0%
Employer benefits	16.8%	17.5%	16.9%	16.6%
Public programs	15.3%	11.1%	16.5%	16.8%
Tort liability	10.5%	7.5%	31.4%	5.3%
Family, friends and other unspecified sources	7.5%	7.5%	5.9%	6.3%

Bibliographic References

ABRAHAM, K.S., Liability Insurance and Accident Prevention: The Evolution of an Idea, *Tort Law, Public Law and Legal Theory*, Working Paper Series, University of Virginia Law School, 2004.

ARROW, K., Uncertainty and the Welfare Economics of Medical Care, *American Economic Review*, 1963, pp. 941–973.

CALABRESI, G., *The Costs of Accidents: A Legal and Economic Analysis*, New Haven: Yale University Press, 1970.

CARROLL, S., *Effects of an Auto-Choice Automobile Insurance Plan on Costs and Premiums*, RAND Corporation Testimony Series, 1997 (published at http://www.rand.org/publications/CT/CT141).

CHELIUS, J.R., The Status and Direction of Workers' Compensation: An Introduction to Current Issues, in: CHELIUS, J.R. (ed.), *Current Issues in Workers' Compensation*, W.E. Upjohn Institute for Employment Research, 1986.

CHELIUS, J.R., Liability for Industrial Accidents: A Comparison of Negligence and Strict Liability Systems, *Journal of Legal Studies*, Vol. 5, 1976, pp. 293–309.

COASE, R., The Problem of Social Cost, *Journal of Law and Economics*, Vol. 3, 1960, pp. 1–44.

COOTER, R. & ULEN, T., *Law & Economics*, Pearson Addison Wesley, fourth edition, 2004.

DANZON, P.M., Compensation for Occupational Disease: Evaluating the Options, *The Journal of Risk and Insurance*, Vol. 54, No. 2, 1987, pp. 263–282.

[119] D.R. Hensler *et al.* (*supra* fn. 11), 108.

DANZON, P.M. & HARRINGTON, S.E., *Rate Regulation of Workers' Compensation Insurance: How Price Controls Increase Costs*, AEI Press, 1998.

DARLING-HAMMOND, L. & KNIESNER, T.J., *The Law and Economics of Workers' Compensation*, RAND Institute for Civil Justice, 1980.

DEWEES, D.N., Economic Incentives for Controlling Industrial Disease: The Asbestos Case, *Journal of Legal Studies*, Vol. 15, 1986, pp. 289–319.

DEWEES, D., DUFF, D. & TREBILCOCK, M., *Exploring the Domain of Accident Law: Taking the Facts Seriously*, Oxford University Press, 1996.

EHRENBERG, R.G., Workers' Compensation, Wages, and the Risk of Injury, in: BURTON, J.F. (ed.), *New Perspectives in Workers' Compensation*, ILR Press, 1988, pp. 71–96.

FAURE, M.G., The Applicability of the Principles of Private Insurance to Social Health Care Insurance, Seen from a Law and Economics Perspective, *The Geneva Papers on Risk and Insurance: Issues and Practice*, Vol. 23, No. 87, 1998, pp. 265–293.

FAURE, M.G., HARTLIEF, T. & PHILIPSEN, N.J., Funding of Personal Injury Litigation and Claims Culture: Evidence from the Netherlands, *Utrecht Law Review*, Vol. 2, No. 2, 2006, pp. 1–21.

FISHBACK, P.V., Liability Rules and Accident Prevention in the Workplace: Empirical Evidence from the Early Twentieth Century, *Journal of Legal Studies*, Vol. 16, 1987, pp. 305–328.

GABEL, J.T.A., Escalating Inefficiency in Workers' Compensation Systems: Is Federal Reform the Answer?, *Workers' Compensation Law Review*, Vol. 22, 2000, pp. 71–124.

GALANTER, M., An Oil Strike in Hell: Contemporary Legends About the Civil Justice System, *Arizona Law Review*, Vol. 40, 1998, pp. 717–752.

GALANTER, M., Real World Torts: An Antidote to Anecdotes, *Maryland Law Review*, Vol. 55, No. 4, 1996, pp. 1093–1160.

GOLDBERG, J.C.P., Twentieth Century Tort Theory, *Georgetown Law Journal*, Vol. 91, 2003, pp. 514–583.

HENSLER, D.R., et al., *Compensation for Accidental Injuries in the United States*, RAND Institute for Civil Justice, 1991.

KLAFF, D.B., Evaluating Work: Enforcing Occupational Safety and Health Standards in the United States, Canada and Sweden, *University of Pennsylvania Journal of Labor & Employment Law*, Vol. 7, 2005, pp. 613–659.

KNIESNER, T.J. & LEETH, J.D., Occupational Safety and Health Administration, in: BOAZ, D. & CRANE, E.H. (eds.), *Cato Handbook for Congress: 105th Congress*, Cato Institute, 1997 (published at http://www.cato.org/pubs/handbook/hb105-36.html).

KÖTZ, H. & SCHÄFER, H.B., Economic Incentives to Accident Prevention: An Empirical Study of the German Sugar Industry, *International Review of Law and Economics*, 1993, Vol. 13, pp. 19–33.

LITTLE, J.W., EATON, T.A. & SMITH, G.R., *Workers' Compensation: Cases and Materials*, Thomson West, fifth edition, 2004.

MCEWIN, R.I., No-Fault Compensation Systems, in: BOUCKAERT, B. & DE GEEST, G. (eds.), *Encyclopedia of Law and Economics, Volume II. Civil Law and Economics*, Cheltenham, Edward Elgar, 2000, pp. 735–763.

MOORE, M.J. & VISCUSI, W.K., *Compensation Mechanisms for Job Risks: Wages, Workers' Compensation, and Product Liability*, Princeton University Press, 1990.

NATIONAL SAFETY COUNCIL, *Injury Facts: 2004 edition*, 2004.

OI, W.Y., On the Economics of Industrial Safety, *Law & Contemporary Problems*, Vol. 38, 1973–1974, pp. 669–699.

PHILIPSEN, N.J., *Regulation of and by Pharmacists in the Netherlands and Belgium: An Economic Approach*, Antwerpen/Groningen: Intersentia, 2003.

POSNER, R.A., *Economic Analysis of Law*, Aspen Publishers, sixth edition, 2003.

REA, S.A., Workmen's Compensation and Occupational Safety Under Imperfect Information, *American Economic Review*, Vol, 71, No. 1, pp. 80–93.

SAKS, M.J., Do We Really Know Anything About the Behavior of the Tort Litigation System – And Why Not?, *University of Pennsylvania Law Review*, Vol. 140, No. 4, 1992, pp. 1147–1292.

SCHWARZ, G.T., Empiricism and Tort Law, *University of Illinois Law Review*, Vol. 2002, No. 4, 2002, pp. 1067–1082.

SHAVELL, S., Strict Liability versus Negligence, *Journal of Legal Studies*, Vol. 9, 1980, pp. 1–25.

SHAVELL, S., Liability for Harm versus Regulation of Safety, *Journal of Legal Studies*, Vol. 13, 1984, pp. 357–374.

VANDOREN, P., Occupational Safety and Health Administration, in: BOAZ, D. & CRANE, E.H. (eds.), *Cato Handbook for Congress: Policy Recommendations for the 108th Congress*, Cato Institute, 2003 (available at http://www.cato.org/pubs/handbook/hb108/hb108-35.pdf).

WORRALL, J.D. & APPEL, D., Some Benefit Issues in Workers' Compensation, in: WORRALL, J.D. & APPEL, D. (eds.), *Workers' Compensation Benefits: Adequacy, Equity and Efficiency*, ILR Press, 1985, pp. 1–18.

Shifts in Compensating Work-Related Injuries and Diseases: Concluding Observations

S. Klosse* and T. Hartlief** (eds.)

1. Confinements

This volume presents a picture of various trends that can be perceived in the way in which damage caused by work-related injuries and diseases is compensated in Germany, England, Belgium and the Netherlands. Even though the contributors to this book have been able to sketch a colourful picture of these trends, it should be kept in mind that this picture only represents a fraction of the developments that took place in this area in the course of time. Undoubtedly, other shifts may have occurred in other legal systems which could not be discussed within the framework of this book. Before making some concluding observations, it is also good to realize that the contributors to this book focussed on uncovering relevant shifts and the motives behind them in the four countries under review and on explaining why sometimes different solutions have been chosen to achieve a certain goal. In view of these confinements, it is rather precarious to make general statements with regard to the effectiveness of the shifts that can be observed and, equally, to come up with normative observations on what would be the optimal compensation scheme. Observations of this kind require additional research material and a deeper insight into the normative goals and the starting points of various legal systems with respect to the compensation of work-related personal damage. These basic notions may be quite different, since they are closely linked with the legal cultural framework which varies form country to country. Further investigation is needed to map and understand the differences stemming from these frameworks; however this would exceed the scope of this book.

2. Common Features

Notwithstanding the limitations just mentioned, it is possible to detect some general trends in the way in which compensation for damage arising from work-related injuries and diseases has been arranged in the four countries un-

1

2

* Prof. S. Klosse, Professor of Employment and Welfare Law, Faculty of Law, University of Maastricht.
** Prof. T. Hartlief, Professor of Private Law, Faculty of Law, University of Maastricht.

der review. The previous chapters show, for example, that three time periods can be distinguished in which shifts cropped up in the way in which damage resulting from work-related injuries and diseases is to be compensated. Interestingly, the circumstances that gave rise to the shifts that have been identified in this book, appear to be by and large the same in Germany, England, Belgium and the Netherlands. At the end of the 19th century, for example, 'social appeasement' was the main driving force behind policy initiatives to rearrange the system. In all the countries under review, the policy change that resulted from these initiatives was meant to alleviate the negative effects of industrialization for workers who often were condemned to live in poverty when they were struck by work-related injuries. To improve this social misery, the existing compensation system was replaced by a system that was more easily accessible and one that strengthened the responsibility of employers to provide workers with adequate compensation in the case of damage arising from work-related injuries.

3   After the Second Word War, economic prosperity and a more strongly felt national solidarity, inspired national governments in the countries under review to further develop and expand the compensation schemes in place. Characteristic of this period is that industrialisation is no longer seen as a process for which employers are financially responsible, but rather as a social fact, the advantages and disadvantages of which have to come down on society as a whole. This notion laid the foundation for a transition towards a state organised overall social security system with strong public law elements, such as mandatory insurance, state regulated minimum and/or maximum levels of compensation and a state controlled administration.

4   In the third period, which started in the late 1970's, considerations of cost containment became predominant in the policy discourse on compensating personal damage. The policy changes that came about in this period would seem to indicate that this development induced policymakers to take a step backwards, for example, by reducing the scope of the compensation system in place, downsizing the level of benefits and/or by shifting costs back from the public to the private sector.

5   In all the countries under review these evolutions are discernable in about the same period, albeit not always to the same extent and with the same outcome. From this it would seem to follow, that decisions to change the compensation system in place are not only provoked by experienced deficiencies in an existing system; common external factors and social circumstances, often combined with pressure and/or effective lobbying of particular interest groups, also play a major role in triggering off policy changes.

3. Different Solutions

6   Even though it is possible to identify a resemblance in the context and the underlying causes for the policy changes that can be observed in the four

countries under review, this does not imply that the policy changes as such have been identical. At the end of the 19th century, for example, the commonly felt need to improve the compensation system in place, gave rise to a shift from civil liability towards a compulsory social insurance scheme for employers in Germany and the Netherlands, whereas in Belgium and England the principle of compulsory insurance was rejected. Instead, strict liability was introduced as a means to offer more adequate compensation facilities to victims of work-related injuries. After the Second World War, the principle of compulsory insurance was accepted in both Belgium and England; however, only in England was this shift accompanied by a formal transition towards social insurance as the prime source of compensation. On top of that, the possibility to take legal action against the employer on the basis of civil liability was preserved in England, whereas in Belgium this possibility was largely put aside. Thus, different instruments were chosen to correct observed inadequacies of the compensation scheme in place.

It is interesting to note, that in spite of the fact that the chosen instruments   7
were unalike, they had some elements in common. After all, they all paved the way for a compensation scheme that was more easily accessible by offering compensation irrespective of fault, thus putting the financial burden of work-related injuries more clearly on the employer. From this it would seem to follow that the pursuit to improve the indemnification of victims of work-related injuries does not necessarily imply a transition from civil liability towards a compulsory insurance system with strong public elements which reveal themselves, for example, in a low threshold of compensation, limited and fixed benefit levels and a maximal spreading of losses, sometimes combined with a differentiation of the insurance premium in function of the risk a company represents with regard to the occurrence of work-related injuries. This is only one of the solutions that can be chosen. History shows us that a more cautious approach is possible as well, for example by remaining closer to the traditional system with strong private law elements.

Illustrative of a more cautious approach is the 'Belgium case'. Characteristic   8
of Belgium is that the ambition to improve the compensation system in place did not bring about 'real' shifts in the paradigm of compensating work-related injuries. Surely, the original Belgium system has been refined over the years, first at the end of the 19th century by introducing strict liability for employers and subsequently after the Second World War by the adoption of the principle of full compensation and the principle of compulsory insurance, while at the same time granting civil immunity to Belgian employers. Thus, in the course of time, public law elements have been introduced in the Belgium system which express themselves for example in fixed levels of compensation and a spreading of losses through compulsory insurance. As a result of this, the Belgium compensation system for work-related injuries came to bear close likeness to the features of the original Dutch and German system, albeit without generating a shift in the paradigm of compensating work-related injuries.

9   Another example of a more cautious approach can be found in the 'German case'. After the shift from civil liability towards a compulsory social insurance system at the end of the 19th century, no 'real' shifts can be observed in the German system. It is of course true that, since the end of the 19th century, the German system has been subject to several alterations so as to further expand the scope of the compensation system in place and to underline the importance of a preventive approach and adequate return to work policies. However, these alterations did not bring about fundamental changes in the German compensation scheme nor in its features. Illustrative in this context is, for one, that compensation on the basis of civil liability is still largely put aside. Furthermore, the German system still allows employers civil immunity in exchange for their duty to take out social insurance for work-related injuries and diseases.

10   Developments that can be identified in the Netherlands and England demonstrate that the pursuit of responding more adequately to the financial needs of victims of work-related injuries can have a different outcome as well. In these countries, this noble aspiration resulted after the Second World War in the decision to integrate the compensation mechanism for work-related injuries in the general social insurance system, accompanied by civil liability as a complementary compensation scheme. In England this decision brought about a fundamental change in the sense that strict liability was replaced by social insurance as a primary source of compensation. Although this primary source was based on industrial preference, it was financed by both employers and employees and the Exchequer. At the same time, insurance for the employer's civil liability became compulsory so as to provide victims of work-related injuries and diseases with the possibility to obtain full compensation.

11   In the Netherlands, the integration of the compensation mechanism for work-related injuries in the general social insurance system implied a second major shift, since, this time, a new mandatory social insurance scheme was introduced which covered the financial risk of both work-related and non work-related personal damage. Work-related injuries and diseases were now seen as a 'risque social' which had to come down on society as a whole and, therefore, should be financed by both the employer and the employee. The introduction of this new scheme produced a fundamental change, not only compared to the features of the other compensation systems under review, but also within the Dutch system itself. This was even more so since civil liability was reintroduced as a complementary compensation scheme in order to enable victims of work-related injuries to obtain full compensation. Unlike in England, this policy change did not simultaneously put employers under the obligation to take out liability insurance.

4.   The Balance Between Public Solidarity and Private Responsibility

12   In recent years, considerations of cost containment have become predominant in the policy discourse on compensating personal damage; a development which has been induced by an increased appeal to social insurance benefits

and, in connection with this, an ongoing debate on the future of the welfare state and the need to keep public expenditure within limits. Interestingly, only in England and the Netherlands, has this development provoked a demand to reconsider the compensation system in place, thereby underlining the importance of finding ways to strike a new balance between public solidarity and private responsibility. The compensation scheme in place would seem to have lost its balance due to the felt necessity to cut down on public expenditure, and, more in particular, on social security benefits, for example by restricting the conditions to claim benefits and reducing the benefit levels. As a result of this, employers have been increasingly faced with claims for compensation on the basis of civil liability, being the complementary compensation scheme in both countries.

Practice shows that this development may have some undesirable effects. For one, lowering the level of income support provided by social insurance is likely to provoke more civil proceedings so as to obtain additional compensation from the employer. Consequently, it is quite possible that this system will lead to more litigation, higher legal and administrative costs and rising private insurance premiums, which in turn can put the liability insurance market under pressure and may cause insecurity with regard to the insurability of work-related injuries and diseases.   13

In addition, this development would seem to induce a warm enthusiasm for various forms of privatisation which aim at confronting employers (and their insurance companies) more directly with the cost of sick-leave and incapacity to work. In England this trend manifests itself, for example, in a new ruling which allows social insurance providers to demand repayment of social security benefits from any person who is held liable for the victim's damages, including the employer. In the Netherlands, employers have been made directly responsible for the financial risk of sick-leave and incapacity to work during the first two years of illness whether it is work-related or not. On top of that, the social insurance scheme for long-term work incapacity due to illness has been reformed. As a result of this, the conditions for entitlement to long-term benefits have been tightened, while simultaneously lowering the benefit level to a percentage of the statutory minimum wage if a remaining work capacity has not been made sufficiently productive after some time.   14

The idea behind this form of privatisation is to give employers and employees a strong incentive to curb the risk of sick-leave and incapacity to work and to strengthen their responsibility to foster a quick return to work. This would seem to imply a shift back from government intervention towards private responsibility, combined with an increasing demand for private insurance and additional compensation based on the employer's liability. Expectations are that this shift from public responsibilities onto the private sector will enhance the development of preventive policies at company level and, consequently, will help to tackle problems related to sick-leave and incapacity to work in a more effective way.   15

16   All the same, the question arises to what extent such a shift will contribute to
      bringing private responsibility and public solidarity into balance. On the basis
      of the findings of the case studies presented in this book, it would seem justifi-
      able to feel doubtful about this. After all, these findings provide evidence that
      shifting the emphasis from a public compensation scheme towards a scheme
      with strong private law elements will only lead to a balanced outcome under
      the condition that essential characteristics of the public system (such as fixed
      levels of compensation set by the government and solidarity elements which
      reveal themselves in low thresholds to enter the system and income related
      premiums based on compulsory insurance) are sufficiently maintained in the
      new private system. If this is not the case, such a shift is likely to produce an
      unsteady balance between private responsibility and public solidarity since, in
      that case, private responsibilities tend to drown out public solidarity due to a
      lack of adequate additional instruments to hold up the solidarity elements
      which are characteristic of a public system.

5.   Some Policy Conclusions

17   As said before, drawing normative statements on the consequences of certain
      policy choices with regard to the way in which the compensation of damage
      arising from work-related injuries and diseases is arranged, goes beyond the
      scope of this book. Yet, this does not alter the fact that the contributions to this
      book provide valuable information on several trends that can be observed in
      this field in the course of time. These trends show that there are different ways
      to achieve a certain goal, such as providing victims of work-related personal
      damage with adequate compensation facilities or enhancing preventive poli-
      cies at company level. As to the choice of a particular instrument to fulfil these
      goals, there would seem to be a certain 'path dependency' which induces the
      legislator to stick to a chosen path that complies with the existing institutional
      structures or with the legal cultural framework in place. It is this framework
      that often eventually determines why, for example, preference is given to a
      more cautious approach towards making fundamental changes in the compen-
      sation scheme in place, which has been the case in Germany and Belgium. It
      may also explain why the Dutch and English legislators adhered to the princi-
      ple of full compensation and, in connection with this, to the preservation of
      the possibility to claim additional compensation from the employer on the ba-
      sis of civil liability and why, unlike in England and the Netherlands, the basic
      assumptions upon which the German and the Belgium schemes are built, have
      never seriously been questioned so far. Since the research questions on which
      the findings of this book are based, not only focus on detecting relevant shifts
      but also on uncovering the motives behind them, it is possible to conclude that
      the legal cultural context is indeed an influential factor when it comes to mak-
      ing decisions on certain policy changes.

18   In spite of this, additional research is necessary to draw firm conclusions on
      the extent to which differences in the actual choices that have been made, re-
      late to differences in the legal cultural context. The findings of this book do

not provide sufficient evidence on this aspect nor do they reveal to what extent differences in the policy choices made, relate to differences in the countervailing power of particular interest groups, such as employer organisations and trade unions. Indeed, there are indications that this countervailing power exists, but on the basis of the findings of this book it is not possible to indicate to what extent certain policy changes can be ascribed to the differences in the legal cultural context or to effective lobbying in favour of a legislative change of particular interest groups. The interrelationship between these two 'powers' in the decision-making process needs further investigation.

Additional information is also needed to gain a deeper insight into the effects 19 of certain policy choices. The contributions to this book display that it is quite hard to obtain relevant data on the basis of which solid conclusions can be drawn on the extent to which the effects hoped for by the legislator, actually occurred. To increase the effectiveness of policy choices which aim at, for example, improving the compensation facilities or curbing the risk of work-related damage, it is of utmost importance that these presumed effects can be analysed and tested on the basis of relevant data. These data should be available or, insofar as they are lacking, should be made available, if necessary by active government intervention.

6. *Epilogue*

At the end of a book which aims at addressing policy changes in the compen- 20 sation of work-related injuries and diseases from the perspective of 'shifts in governance', it would seem fair to conclude that all in all this has proved to be a valuable exercise. For one, this approach would seem to provide a useful analytic tool which helps to uncover fundamental changes in the legal basis to bear the damage arising from work-related injuries and diseases. In addition, it helps to focus more clearly on the underlying question of whether this sort of damage should be seen as a result of an industrial process for which the employer is primarily responsible or rather as a social fact, the consequences of which have to come down to society as a whole and therefore should be born by the employer, the employee and the state. Moreover, the shifts approach underlines the importance of distinguishing between the motives for certain policy choices that are formally presented by the legislator and the influence of other factors that, for example, relate to pressure and/or successful lobbying efforts of particular interest groups. This may lead to a growing awareness that legislative initiatives are often inspired by a variety of different reasons and therefore give rise to a certain precaution, especially in performing an ex post analysis of various shifts. On top of that, this approach helps to keep in mind that shifts are not always taking place for reasons of public interests but may serve private interests as well. In turn, this may clarify that policy analyses should not be confined to the question of whether public or private governance is preferable, but rather include the possibility of examining the possible advantages of a mix between these two basic forms of governance. The contributions to this book show that the shifts approach does not necessarily result in firm conclusions at the normative level, for example, on what the best mix would be in terms of compensating or preventing work-related personal damage. To make solid policy recommendations on this particular aspect, an additional analysis is needed. Seen from this perspective the research conducted within the framework of this volume can be regarded as a fruitful first step that lays the foundation for further research on the fascinating theme of this book.

Index

Publications

Principles of European Tort Law

Volume 1: *The Limits of Liability: Keeping the Floodgates Shut.*
Edited by Jaap Spier.
Kluwer Law International, The Hague. Hardbound.
ISBN 90-411-0169-1. 1996, 162 pp.

Volume 2: *The Limits of Expanding Liability. Eight Fundamental Cases in a Comparative Perspective.*
Edited by Jaap Spier.
Kluwer Law International, The Hague. Hardbound.
ISBN 90-411-0581-6. 1998, 244 pp.

Volume 3: *Unification of Tort Law: Wrongfulness.*
Edited by Helmut Koziol.
Kluwer Law International, The Hague. Hardbound.
ISBN 90-411-1019-4. 1998, 144 pp.

Volume 4: *Unification of Tort Law: Causation.*
Edited by Jaap Spier.
Kluwer Law International, The Hague. Hardbound.
ISBN 90-411-1325-8. 2000, 161 pp.

Volume 5: *Unification of Tort Law: Damages.*
Edited by Ulrich Magnus.
Kluwer Law International, The Hague. Hardbound.
ISBN 90-411-1481-5. 2001, 225 pp.

Volume 6: *Unification of Tort Law: Strict Liability.*
Edited by Bernhard A. Koch and Helmut Koziol.
Kluwer Law International, The Hague. Hardbound.
ISBN 90-411-1705-9. 2002, 444 pp.

Volume 7: *Unification of Tort Law: Liability for Damage Caused by Others.*
Edited by Jaap Spier.
Kluwer Law International, The Hague. Hardbound.
ISBN 90-411-2185-4. 2003, 335 pp.

Volume 8: *Unification of Tort Law: Contributory Negligence.*
Edited by Ulrich Magnus and Miquel Martín-Casals.
Kluwer Law International, The Hague. Hardbound.
ISBN 90-411-2220-6. 2004, 300 pp.

Volume 9: *Unification of Tort Law: Multiple Tortfeasors.*
Edited by W.V. Horton Rogers.
Kluwer Law International, The Hague. Hardbound.
ISBN 90-411-2319-9. 2004, 313 pp.

Volume 10: *Unification of Tort Law: Fault.*
Edited by Pierre Widmer.
Kluwer Law International, The Hague. Hardbound.
ISBN 90-411-2098-X. 2005, 393 pp.

Tort and Insurance Law

Volume 1: *Cases on Medical Malpractice in a Comparative Perspective.*
Edited by Michael Faure and Helmut Koziol.
Springer, Vienna/New York.
Softcover. ISBN 3-211-83595-4. 2001, 331 pp.

Volume 2: *Damages for Non-Pecuniary Loss in a Comparative Perspective.*
Edited by W.V. Horton Rogers.
Springer, Vienna/New York.
Softcover. ISBN 3-211-83602-0. 2001, 318 pp.

Volume 3: *The Impact of Social Security on Tort Law.*
Edited by Ulrich Magnus.
Springer, Vienna/New York.
Softcover. ISBN 3-211-83795-7. 2003, 312 pp.

Volume 4: *Compensation for Personal Injury in a Comparative Perspective.*
Edited by Bernhard A. Koch and Helmut Koziol.
Springer, Vienna/New York.
Softcover. ISBN 3-211-83791-4. 2003, 501 pp.

Volume 5: *Deterrence, Insurability and Compensation in Environmental Liability. Future Developments in the European Union.*
Edited by Michael Faure.
Springer, Vienna/New York.
Softcover. ISBN 3-211-83863-5. 2003, 405 pp.

Volume 6: *Der Ersatz frustrierter Aufwendungen. Vermögens- und Nichtvermögensschaden im österrei-chischen und deutschen Recht.*
By Thomas Schobel.
Springer, Vienna/New York.
Softcover. ISBN 3-211-83877-5. 2003, 342 pp.

Volume 7: *Liability for and Insurability of Biomedical Research with Human Subjects in a Comparative Perspective.*
Edited by Jos Dute, Michael G. Faure and Helmut Koziol.
Springer, Vienna/New York.
Softcover. ISBN 3-211-20098-3. 2004, 445 pp.

Volume 8: *No-Fault Compensation in the Health Care Sector.*
Edited by Jos Dute, Michael G. Faure, Helmut Koziol.
Springer, Vienna/New York.
Softcover. ISBN 3-211-20799-6. 2004, 492 pp.

Volume 9: *Pure Economic Loss.*
Edited by Willem H. van Boom, Helmut Koziol and Christian A. Witting.
Springer, Vienna/New York.
Softcover. ISBN 3-211-00514-5.
2003, 214 pp.

Volume 10: *Liber Amicorum Pierre Widmer.*
Edited by Helmut Koziol and
Jaap Spier.
Springer, Vienna/New York.
Softcover. ISBN 3-211-00522-6.
2003, 376 pp.

Volume 11: *Terrorism, Tort Law and Insurance. A Comparative Survey.*
Edited by Bernhard A. Koch.
Springer, Vienna/New York.
Softcover. ISBN 3-211-01867-0.
2004, 313 pp.

Volume 12: *Abschlussprüfer.*
Haftung und Versicherung.
Edited by Helmut Koziol and
Walter Doralt.
Springer, Vienna/New York.
Softcover. ISBN 3-211-20800-3.
2004, 180 pp.

Volume 13: *Persönlichkeitsschutz gegenüber Massenmedien/The Protection of Personality Rights against Invasions by Mass Media.*
Edited by Helmut Koziol and
Alexander Warzilek.
Springer, Vienna/New York.
Softcover. ISBN 3-211-23835-2.
2005, 713 pp.

Volume 14: *Financial Compensation for Victims of Catastrophes.*
Edited by Michael Faure and
Ton Hartlief.
Springer, Vienna/New York.
Softcover. ISBN 3-211-24481-6.
2006, 466 pp.

Volume 15: *Entwurf eines neuen österreichischen Schadenersatzrechts.*
Edited by Irmgard Griss,
Georg Kathrein and Helmut Koziol.
Springer, Vienna/New York.
Softcover. ISBN 3-211-30827-X.
2006, 146 pp.

Volume 16: *Tort Law and Liability Insurance.*
Edited by Gerhard Wagner.
Springer, Vienna/New York.
Softcover. ISBN 3-211-24482-4.
2005, 361 pp.

Volume 17: *Children in Tort Law.*
Part I: Children as Tortfeasors.
Edited by Miquel Martín-Casals.
Springer, Vienna/New York.
Softcover. ISBN 3-211-24480-8.
2006, 476 pp.

Volume 18: *Children in Tort Law.*
Part II: Children as Victims.
Edited by Miquel Martín-Casals.
Springer, Vienna/New York.
Softcover. ISBN 3-211-31130-0.
2007, 320 pp.

Volume 19: *Tort and Regulatory Law.*
Edited by Willem H. van Boom,
Meinhard Lukas and Christa
Kissling.
Springer, Vienna/New York.
Hardcover. ISBN 978-3-211-31133-20.
2007, approx. 500 pp. (forthcoming).

Volume 20: *Shifts in Compensating Work-Related Injuries and Diseases.*
Edited by Saskia Klosse and
Ton Hartlief.
Springer, Vienna/New York.
Hardcover. ISBN 978-3-211-71555-0.
2007, 236 pp.

Volume 21: *Shifts in Compensation for Environmental Damage.*
Edited by Michael Faure and Albert Verheij.
Springer, Vienna/New York.
Hardcover. ISBN 978-3-211-71551-2.
2007, 338 pp.

European Tort Law Yearbook

European Tort Law 2001.
Edited by Helmut Koziol and Barbara C. Steininger.
Springer, Vienna/New York.
Softcover. ISBN 3-211-83824-4.
2002, 571 pp.

European Tort Law 2002.
Edited by Helmut Koziol and Barbara C. Steininger.
Springer, Vienna/New York.
Softcover. ISBN 3-211-00486-6.
2003, 596 pp.

European Tort Law 2003.
Edited by Helmut Koziol and Barbara C. Steininger.
Springer, Vienna/New York.
Softcover. ISBN 3-211-21033-4.
2004, 493 pp.

Principles of European Tort Law

Text and Commentary.
Edited by the European Group on Tort Law.
Springer, Vienna/New York.
Softcover. ISBN 3-211-23084-X.
2005, 282 pp.

Digest of European Tort Law

Volume 1: *Essential Cases on Natural Causation.*
Edited by Bénédict Winiger, Helmut Koziol, Bernhard A. Koch and Reinhard Zimmermann.
Springer, Vienna/New York.
Hardcover. ISBN 978-3-211-36957-9.
2007, 632 pp.

Volume 22: *Shifts in Compensation Between Private and Public Systems.*
Edited by Willem H. van Boom and Michael Faure.
Springer, Vienna/New York.
Hardcover. ISBN 978-3-211-71553-6.
2007, 246 pp.

European Tort Law 2004.
Edited by Helmut Koziol and Barbara C. Steininger.
Springer, Vienna/New York.
Softcover. ISBN 3-211-24479-4.
2005, 674 pp.

European Tort Law 2005.
Edited by Helmut Koziol and Barbara C. Steininger.
Springer, Vienna/New York.
Softcover. ISBN 3-211-31135-1.
2006, 711 pp.

European Tort Law 2006.
Edited by Helmut Koziol and Barbara C. Steininger.
Springer, Vienna/New York.
Hardcover. ISBN 978-3-211-70937-5.
2007, approx. 620 pp. (forthcoming).

Springer and the Environment

WE AT SPRINGER FIRMLY BELIEVE THAT AN INTER-national science publisher has a special obligation to the environment, and our corporate policies consistently reflect this conviction.

WE ALSO EXPECT OUR BUSINESS PARTNERS – PRINTERS, paper mills, packaging manufacturers, etc. – to commit themselves to using environmentally friendly materials and production processes.

THE PAPER IN THIS BOOK IS MADE FROM NO-CHLORINE pulp and is acid free, in conformance with international standards for paper permanency.